The FUR FARMS of ALASKA

The FUR FARMS of ALASKA

Two Centuries of History and a Forgotten Stampede

Sarah Crawford Isto

University of Alaska Press
Fairbanks

© 2012 University of Alaska Press
All rights reserved
University of Alaska Press
P.O. Box 756240
Fairbanks, AK 99775-6240

ISBN 978-1-60223-171-9 (paper); 978-1-60223-172-6 (e-book)

Library of Congress Cataloging-in-Publication Data
Isto, Sarah Crawford, 1942–
 The fur farms of Alaska : two centuries of history and a forgotten stampede / Sarah Crawford Isto.
   p. cm.
 Includes bibliographical references and index.
  ISBN 978-1-60223-171-9 (pbk. : alk. paper) — ISBN 978-1-60223-172-6 (e-book)
 1. Fur farming—Alaska—History. 2. Farms—Alaska—History. I. Title.
 SF403.4.A4I78 2012
 636.9709798—dc23
   2011033336

Cover design by Publishers' Design and Production Services, Inc.

Cover photo: Live muskrats being shipped from Kodiak area to Sand Point, Alaska. Courtesy of Mary Graves Zahn.

This publication was printed on acid-free paper that meets the minimum requirements for ANSI / NISO Z39.48–1992 (R2002) (Permanence of Paper for Printed Library Materials).

Printed in the United States on recycled paper.

CONTENTS

	Preface	*ix*
	Acknowledgments	*xi*
	Introduction	*1*
One	**The Russian Period, 1749–1866**	7
	Planting Foxes on Aleutian Islands	
	Siberian Foxes on Attu Island	12
	Native Silver Fox Transplants	14
	The Pribilof Islands: A Blue Fox Enclave	16
	Russian-American Company Fox Farms	18
	Aleut Warning: Foxes Destroying Bird Colonies	20
Two	**The Pioneer Period, 1867–1909**	23
	Americans Reinvent Alaska Fur Farming	
	Government Fox Sales and Fur Farm Leases	24
	The Semidi Fox Propagation Company on Long Island	27
	James Judge and the Pribilof Fox Farm	30
	Thomas Vesey Smith on Middleton Island	33
	Silver Foxes in Pens	36
	Charles Heideman: Fox Farm Promoter at Copper Center	40
Three	**Prewar Expansion and WWI, 1910–1918**	43
	Conservationist Support for Alaska Fur Farms	
	Forest Service Fur Farm Permits	44
	The Biological Survey and New Farmed Species	46
	Bolshanin, Applegate, and the Aleuts	49

Contents

The Territory of Alaska and the Eagle Bounty	54
Farm Failures: The Semidi Fox Propagation Company	57
W. J. Erskine and the Kodiak Fur Farm on Long Island	58
Bureau of Fisheries: New Managers on the Pribilofs	62
Wartime Losses and Profits	63

Four — The Fur Farm Rush, 1919–1924 — 65
"A Stampede to Take Up Islands"

Fur Farms throughout Alaska	68
Taxes, Merchants, and the Seattle Fur Exchange	73
Poaching and the Fox Branding Law	79
Alaska's Fur Farm Associations	81
Women Fur Farmers	84
Domesticating Foxes	89
Saturated Fur Markets, Animal Disease, and the Price of Feed	92

Five — The Peak Years, 1925–1929 — 95
A Territorial Veterinarian and 700 Fur Farms

The Alaska Fur Farm Leasing Act	96
Basal Parker: Veterinarian and Fur Farmer	97
The National Experimental Fur Farm	100
Earl Graves: First Territorial Veterinarian for the Fur Farms	101
Exporting Live Breeders from Alaska	105
Disease, Negligence, and Moonshine in the Islands	108
The Alaska Game Commission and Fur Farm Statistics	110

Six — The Great Depression 1930–1940 — 115
Depression Years and Alaska's Experimental Fur Farm

Closing Farms and Changing Fashions	118
Government Aid to Failing Farms	119
Jule Loftus: Last Territorial Veterinarian for the Fur Farms	120
Fur Farming in the Depth of the Depression	125

	The Alaska Experimental Fur Farm	128
	Territorial Fur Farms that Survived the Depression	132
	European War and American Fur Markets	139
Seven	**World War II, 1941–1945**	**141**
	A Nonessential Industry in a War Zone	
	A Farm Lost to Internment	142
	Aleutian Fox Farms under Double Attack	144
	Pribilof Fox Farming Suspended	147
	Furs in the War Effort	148
	Government Controls on Furriers and Fur Sales	149
	James Leekley at the Alaska Experimental Fur Farm	150
	The War's Legacy: Sixty Surviving Alaska Fur Farms	154
Eight	**Post-War Hopes and Decades of Decline, 1946–2000**	**157**
	Cold War, Oil Boom, and the Demise of Alaska Fur Farming	
	The Aleutians: Fox Farming Ends, Fox Eradication Begins	158
	Pribilof Fox Farming after WWII	160
	Southeast Alaska: Native Rights and New Forest Service Policies	161
	Post-War Culture and Fashion Changes	164
	Fur Farms under Statehood	168
	Alaska's Experimental Fur Farm in the Jet Age	170
	The Final Decades: Mutation Mink, Silver Fox, and PETA	173
	Alaska Fur Farming: Dreams and Realities	176
	Endnotes	*179*
	Reference List	*211*
	Index	*221*

PREFACE

Several years ago when I was writing a memoir about my family's early years in Alaska, a cousin mailed to me a stack of unorganized, untitled essays. The essays had been written by my "shirttail uncle," Jule B. Loftus, for a class at his retirement home in Oregon. Uncle Jule, the husband of my mother's first cousin, had been a favorite of mine. He loved practical jokes, kids, and animals. The page on the top of his stack of papers began, "In the spring of 1930 came an opportunity to become Territorial Veterinarian for Alaska and I accepted without hesitation. . . . My duties were . . . visiting all the fur farmers and rendering assistance." Jule went on to describe his adventures with a string of Alaska fur farmers whose characters remained impressed on his mind fifty years later. His ten years as territorial veterinarian were a part of my uncle's life about which I knew nothing. How many fur farmers could there have been in Alaska in 1930?

With a little digging in archives, I found a list of licensed fur farmers and an agricultural survey from the 1929 census. The astonishing answer to my question was that in 1930, when the population of the territory was 60,000, Alaska had more than 600 fur farms. Most of these farms were on islands in the Aleutians and in southeast Alaska, but others clustered on the Kenai and Seward Peninsulas, lined the route of the Alaska Railroad, or bordered interior Alaska's great rivers. There was even a farm about 100 miles east of Point Barrow, the northernmost tip the United States. Alaska fur farmers were raising—or trying to raise—blue fox, silver fox, white fox, mink, beaver, muskrat, marten, otter, lynx, chinchilla rabbits, raccoon, European fitch, and even skunks.

In 1930, fur farming was—as it had been for the previous fifteen years and would be for the next ten years—the third largest industry in Alaska. The year that Jule Loftus started his job as territorial veterinarian, 9,000 blue and silver fox pelts from Alaska farms sold for a total of $609,000 (about $8 million in today's dollars).[1] A majority of the 27,800 mink pelts from Alaska that sold in that same year for $236,000 also came from fur farms. And the largest fur farm in the territory was not even included in these statistics. On the Pribilof Islands, the U.S. federal government was nurturing blue foxes. During the 1920s, sales of Pribilof fox pelts contributed an average of $53,000 per year to the U.S. Treasury.

I also discovered that pelts were not the only product sold by Alaska fur farmers. In 1929, the Bureau of Customs' records show that more than 3,000 live foxes and mink (total value approximately $160,000) were shipped out of Alaska. The exported animals were breeding pairs sold to fur farmers in the United States and northern Europe. Additional live breeders were sold within the territory, but no government agency recorded these sales.

The year 1930 was the midpoint of the American experience with fur farming in Alaska. However, the Russians had pioneered fox farming in the Aleutians more than a hundred years before the United States purchased Alaska. I was curious not only about my uncle's experiences but also about the history of Alaska fur farming before and after his time. The state library had no books on the subject, and I soon realized that if I wanted to follow the story, I would have to piece it together from government reports, trade journals, newspapers, magazines, and memoirs. These were stored in an assortment of libraries, archives, museums, personal photo albums, and the memories of a few living people.

The longer I worked to put this information together, the better I understood why we Alaskans know so little about this colorful episode in our state's history. Fur faming was a transient industry, largely the province of small owners. It is not surprising that the remaining records are scattered and half-buried in the midst of larger collections, nor is it surprising that no one has made an effort to compile this history. Not only are sources challenging to locate but also for some of us—in these days of polypro parkas and animal rights—our history of raising mammals for pelts is discomforting and best left alone.

But fur farming in Alaska is an engaging piece of history replete with political ironies, unrealistic hopes, and impressive innovations. It encompasses fashions and scams, economic freedom and exploitation, success and failure. Most of all, it is a version of a traditional Alaskan saga—the quest by men and women on the margin of wilderness to make an independent living from a renewable natural resource.

ACKNOWLEDGMENTS

I could not have written this book without the dozens of people who helped me find documents, directed me to new material, and supplied me with photos. I am particularly indebted to librarians and archivists at the Alaska State Library (especially Jim Simard and staff), the Alaska State Archives; the Smithsonian Institution Archives, the University of Washington Archives, the National Archives in Anchorage, the University of Alaska Fairbanks Archives (especially Rose Speranza), the Baranov Museum in Kodiak, and the Anchorage Museum at Rasmuson Center. I also received assistance from cultural resource specialists in the U.S. Forest Service (Juneau and Petersburg) and in the National Park Service (Kenai Fjords and Glacier Bay).

In addition, I greatly appreciate the generous help of individuals. Among them are my cousin, Jule H. Loftus, who introduced me to his uncle's manuscript; Sarah McGowan, who sent a box of her files on fur farming in the Aleutians; Larry D. Roberts, who gave me access to his database on southeast Alaska fur farms; Dee Longenbaugh, who pointed out lesser-known books with Alaska fur farming references; and Gary Dederer, who shared his family's photos of the Seattle Fur Exchange. The names of many others can be found in the endnotes at the back of the book and in photo citations throughout the chapters.

In particular I want to express my gratitude to two people who offered unfailing support from the inception of the manuscript. The first is Mary Graves Zahn, who shared with me the well-labeled photograph collection of her father, Territorial Veterinarian Earl F. Graves. The second is my husband Gordon Harrison, whose map work, encouraging advice, and thoughtful reading were invaluable.

INTRODUCTION

In the mid-eighteenth century when Vitus Bering left Siberia and sailed eastward into the North Pacific, he found what he was looking for—a nearby extension of North America that was rich in furs. Russian entrepreneurs were quick to respond to this discovery. They formed companies, assembled boats, and rushed to collect the bounty. These early merchants understood that it was important to be among the first because, under pressure of intense hunting, the animals they sought were sure to decline. The Russians had learned this lesson in the Kurile Islands, and they were about to relearn it in the Commander Islands. Nature could not keep up with the heavy demand for fur garments in China and Northern Europe.

For thousands of years, nature had also been unable to keep up with civilization's demand for leather, milk, and bread. Humans responded by domesticating cattle and growing wheat. Trappers in various places had similarly tried to augment nature by feeding or protecting wild furbearers, but little had come of these efforts. An early Russian explorer, captain of a hunting boat headed for Alaska, considered the problem of this inevitable decline in furbearers under the onslaught of unregulated hunting. In 1749, just seven years after the return of Bering's expedition, he responded by establishing Alaska's first fur farm.

Alaska continued to have active fur farms until at least 1993, but the term *fur farm* included a variety of practices. The simplest was a pattern begun by the Russians and continued by Aleuts and early Americans. The farmer collected some breeding foxes, released them onto an island where no foxes lived and gave them time to multiply. A year or two later, the farmer would return in midwinter when the foxes' fur was thickest. He would set leghold traps to catch as many as he could during the several weeks that the fur remained prime. The sea acted as a fence for his farm; beach fauna and migrating birds provided the food.

Introduction

A second and more complex version of island farming required a regular caretaker who prepared and delivered food to feed houses built for free-ranging foxes on the island. In winter, these houses were used to trap the foxes uninjured so that the best animals could be marked, counted, and saved as breeders. When demand for breeders was high, most of the farmer's sales would consist of live animals, which brought far higher prices than pelts. When demand for breeders was low, all but a stable core of animals were killed for their skins.

Alaska's island fox farms were an innovation that received attention in the United States and across the Atlantic. At first, the island method seemed impressively successful, but as years passed, farmers ran into trouble. Poachers, hookworms, diminishing natural food, and expensive transportation precipitated many failures.

In response, during the early twentieth century a more sophisticated model of fur farming began to take hold in Alaska. It followed the pattern used to raise conventional barnyard animals. The farmer kept his mink, foxes, or other furbearers in pens. Breeders with particular characteristics were mated, and records of lineage were often kept. Food, vitamins, and veterinary care were provided. Many farmers also made a significant effort to domesticate their animals. Farms with pens ranged from large operations with hundred-foot-long animal sheds and several employees to part-time "backyard farms," with three or four foxes in an outdoor enclosure or a half-dozen chinchilla rabbits caged in a back room. The latter were rarely included in official reports and statistics, but anecdotal evidence suggests that raising a few furbearers as a sideline was widespread in Alaska during the years before WWII.

Like all industries, fur farming developed a special language. Regardless of whether animals were penned or running free, furbearers of a single species on a farm were called a *herd*. One farmer might have a herd of thirty mink, whose offspring were called *kits* or *kittens*. His neighbor might be raising a fox herd that included twenty adults and a hundred *pups*. A few writers tried to distinguish between methods of raising furbearers by using the term *fur ranch* for free-ranging animals and *fur farm* for animals in pens, but this distinction never became a part of common language. Alaska Mink Ranchers of Cordova raised their animals in pens, while the Sunset Island Blue

Introduction

Fox Farm sold breeders from a herd that ran nearly wild on an island. In this book, the generic term *fur farm* is used to describe both.

In addition to its own terminology, the industry had specialized journals, government fur experiment stations, and a collection of farm bulletins on raising various species. In Alaska, many fur farmers were innovative individuals sustained by a patchwork of rural occupations, some of which—fishing, guiding, logging, and gardening—were older than raising furbearers. Fur farming was the modern approach to harvesting furs—superior to trapping in the quality of the fur, the humane treatment of animals, and the conservation of wild furbearers. It was an industry that fit well with a general optimism about progress and about human control of nature that was common in the late nineteenth and early twentieth centuries. Although Alaska was one of the earliest locations to have fur farms, it was far from alone in this endeavor. By the 1930s several northern tier states had surpassed Alaska in farmed fur production. Today Denmark, Finland, China, Poland, and the Netherlands all produce more farmed pelts than the remaining handful of fur farming regions in the northern United States.

Alaska historically had many assets for raising furbearers: a cold climate that produced thick furs, large areas of land available for farms, sympathetic federal and regional governments, and a populace attracted to rural life and pioneer industry. It also had significant drawbacks: expensive transportation, high food and energy costs, and almost no veterinarians. But the major factor in the balance between success and failure of fur farms was the fur market. During the era of Alaska fur farming, this market was buffeted by the vagaries of fashion, two world wars, the Great Depression, the development of synthetic cold-weather clothing, and animal rights activism.

In Alaska, the natural assets and drawbacks of location combined with a fickle market to create a constantly changing environment for fur farmers. Like many other Alaska industries, fur farming followed a roller-coaster pattern of booms and busts. This book is an attempt to chronicle the origins, rise, striking peak, and slow demise of the industry—and to tell the story of the Alaskans who responded to its challenges.

Bering Sea traversed by Russian ships sailing between the Kamchatka Peninsula and Aleutian Islands. *Drawing courtesy of Gordon Harrison.*

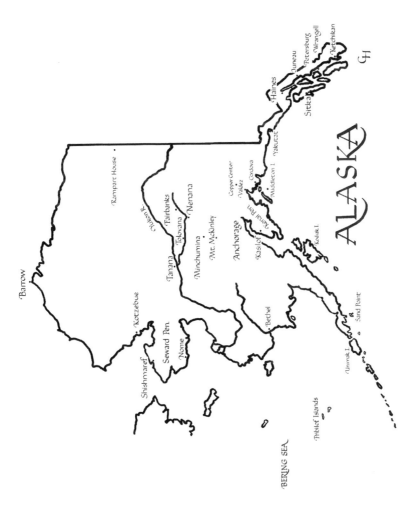

Alaska map with some of the place names mentioned in the book. *Drawing courtesy of Gordon Harrison.*

CHAPTER ONE

The Russian Period
1749–1866

Planting Foxes on Aleutian Islands

In 1749, the Russian fur seeker Andrean Tolstykh and his crew sailed west from Siberia's Kamchatka Peninsula into the easterly winds of autumn. They were steering toward Russia's new territory, the Aleutian Islands. After two months of rough seas, they reached the halfway point of their journey and stopped to winter their ship, the *Sv. Ioann (St. John)* at the Commander Islands.[1] The crew hauled the ship ashore, and for six months they hunted, fished, trapped, and waited out the storms. When the westerly winds of spring arrived, the *Sv. Ioann* set sail again with new provisions, a few furs, and a living cargo whose descendents would affect Alaska's fur trade for two centuries and alter the ecology of some Aleutian Islands forever. In his ship's hold, Andrean Tolstykh nurtured a litter of Siberian arctic foxes.[2]

Tolstykh was a merchant from Selenginsk, a town in southern Siberia near Lake Baikal. Selenginsk was a rest stop for Russian traders and their closely guarded pack horses loaded with goods for China. South of town a road wound 100 miles to the trading stronghold of Kiakhta. This Russian woodblock fortress, population 3,000, stood at the single opening in China's closed northern border. A mere 150 yards to the south, Fort Mai-mai-cheng, surrounded by a moat and camel corrals, housed 1,200 Chinese merchants.[3] A treaty allowed traders to cross between the forts to inspect the merchandise and barter in the common trade language, Mongolian. Most trades were for each country's premium product: Chinese teas and Russian furs. Russia filled its samovars, and Chinese tailors lined cloaks, hats, mitts, and boots with fur. In Beijing, Russian sea otter pelts brought particularly high prices. But by the 1730s, the over-hunted sea otters of coastal Siberia and the Kurile Islands were disappearing. The Russians had even been reduced to purchasing assorted furs in London and transporting them across the continent to meet the demand from China.

Russia needed furs not only for export but also for the domestic market. Royalty and the wealthy of St. Petersburg demanded sable, ermine, and silver fox for ceremony and style.[4] Less exalted Russians needed utilitarian hats, coats, capes, and boot linings. In Siberia, furs even functioned as a monetary system when rubles were scarce. The continual search for fur had impelled Russia on its 200-year expansion eastward toward the Pacific.

Russian merchants along the route from Kamchatka to Kiakhta must have been elated when Vitus Bering's expedition returned in 1742 to report that the northern coast of America was nearby and that its waters teemed with fur seals and sea otters. By the following year, the first commercial vessel had already been constructed in Kamchatka and had set off for a two-year voyage to the Aleutians. The fur take from this first venture is not recorded, but on its second voyage the same ship unloaded a wealth of pelts: 2,240 blue fox, 1,990 fur seal, and 1,670 sea otter. The value in port was 112,220 rubles (more than $700,000 in today's currency).[5] A rush of fur-seeking expeditions followed.

Andrean Tolstykh made his way to Kamchatka early in the rush. He was captain of the fourth commercial ship to leave harbor at the mouth of the Kamchatka River in search of New World furs. But instead of sailing to the Aleutians on his first voyage, he headed southeast in a vain attempt to find the mythical, fur-rich DeGama Land that appeared on a 1733 chart of the North Pacific. Tolstykh's hopes of finding DeGama Land had been raised by notes entered in Bering's log by naturalist Georg Steller. Steller described sighting birds and seaweed that suggested there was land somewhere south of the western Aleutians.[6] Tolstykh discovered no new islands on this first exploratory voyage and brought back few furs. But the inland-born merchant (who may have previously sailed on giant Lake Baikal) found he had an affinity for the ocean. He would soon develop a reputation as a skilled Bering Sea navigator. Eventually he would be credited with the discovery of the Andreanof Islands, named in his honor.

On his second voyage, during which he picked up the litter of arctic foxes, Tolstykh was not only captain and navigator of the *Sv Ioann*, he was also an investor. In the early days of Russian-America, a group of merchants and perhaps a ship builder would form a short-term company for a specific voyage. If the vessel survived storms, uncharted rocks, disorienting fogs, and occasional Aleut hostility to return two or three years later filled with pelts, everyone profited: the investors, the crew working on shares, the government collecting furs for tribute and taxes,[7] and even the church, which received a share in gratitude for the voyage's heaven-ordained survival. The official counting of pelts at the wharf was watched closely by a crowd of interested parties.

This was high-risk venture capitalism. The *Sv. Ioann* was flimsy by modern standards. She was a *shitik*, or "sewn one," built largely from local materials since the cost of importing lumber, nails, and anchors by pack train from Yakutsk was prohibitive. *Shitiki* were made from hand-hewn, green planks that were lashed together with leather thongs or willow wands, which swelled and tightened when wet. Caulking was moss, and anchors were local stones lashed to a wooden framework. The shallow keels and nearly flat bottoms allowed these square-rigged, single-decked ships to be hauled onto a beach for the winter.[8] On board, the *shitiki* carried skin boats for hunting marine mammals and for going ashore. The *Sv. Ioann*'s dimensions are unknown, but a similar ship of the same era measured 42 × 15 feet and carried forty-four men in very close quarters. Most of the crew were hunters who doubled as sailors.[9] The typical navigator's instruments were a telescope, a compass, and rough charts drawn by previous voyagers.

In the first ten years of the fur rush before nailed ships came into common use, as few as four out of the six ships that left Kamchatka during a typical year returned.[10] But not every shipwreck was fatal. The flexible *shitiki* handled deep-water storms well. When these ships came to grief, it was on near-shore rocks. Crew members, sometimes hauling bundles of furs, were frequently able to escape to shore in their skin boats. On shore they dug shelters and awaited rescue or, when possible, used *shitiki* construction methods to lash the wreckage into a makeshift vessel for their return home.[11]

Although sea otter pelts were the most valuable cargo in ships returning from Alaska, fox pelts would eventually account for the greatest volume.[12] The fashions of the time favored long-haired, dark fur. The small, short-eared foxes carried in the *Sv. Ioann*'s hold had dark hair in summer as well as winter. They were a relatively rare color for their species, *Alopex lagopus*.[13] *A. lagopus*, commonly called the arctic fox, is native to circumpolar tundra regions, including islands where sea ice forms a solid bridge to the mainland for winter travel. Most of these arctic foxes are tawny brown in summer and white in winter, but the coat of the melanistic variant—the "blue" arctic fox—is sooty brown in summer and gray-blue in winter. Although the dark color is the result of a dominant gene, most arctic foxes are white in winter. The advantage of camouflage for hunting on tundra or sea ice strongly favors the recessive white coloration.

Like all observant trappers, Tolstykh understood that coat color was the result of breeding. White parents produced only white pups. Blue parents occasionally bore a white or smoky-white pup, but most of their offspring were blue. And blue was a far more valuable color in the fur market than white. In Kamchatka, a blue fox pelt sold for fifteen to forty rubles, whereas a white fox

Arctic fox, white color phase, winter beach scene on Pribilof Islands.
Alaska State Library, Michael Z. Vinokouroff Photo Collection, P243–176.

The Russian Period 1749–1866

Arctic fox, blue color phase, winter scene on Pribilof Islands.
Alaska State Library, Michael Z. Vinokouroff Photo Collection, P243–175.

pelt sold for one to three rubles.[14] Unfortunately for trappers, in most arctic regions, the white foxes far outnumbered the blue.

However, the Commander Islands, where Andrean Tolstykh acquired his blue foxes, were isolated from the Asian mainland except during rare, very cold winters. Most winters the ice pack did not extend far enough to form a solid bridge to the islands. The inbred Commander Island foxes, winter hunters of beaches rather than expanses of ice and tundra, were predominantly blue. In the seven years since Bering's return, these dark foxes had been subject to heavy trapping pressure from over-wintering ships and were at risk of becoming too few for commercial exploitation.[15]

Tolstykh was to make a career of Bering Sea furs. Even in the first years of the rush, he took the long view. He decided that the remedy for over-trapping on the Commander Islands was to establish blue fox colonies on isolated islands elsewhere. Islands for these new fox populations had to meet certain criteria. They had to be free of native white foxes that would dilute the dark color of future generations. They needed to be far enough from the mainland or other islands so that transplanted animals could not swim away during summer or walk away on an ice bridge during winter. Moreover, they needed to possess a generous supply of natural food. Tolstykh looked to the western Aleutians. He had been told that their beaches were richly strewn with shellfish and that the islands were free not only of foxes but of all land mammals save scattered bands of Aleuts.[16] Although these treeless islands lacked mice and ground squirrels, their rocky cliffs and uplands were summer home to ideal fox prey—ground-nesting migratory birds and their eggs. The islands represented new, unused Russian countryside awaiting development.

Siberian Foxes on Attu Island

Tolstykh's long-term plan required the cooperation of the Aleuts living on the island of Attu. In 1745, the first encounter between the Russians and the Aleuts of the "Near Islands" (nearest Siberia) had ended badly when an Agattu Native was shot and wounded. The Russian adventurers then moved thirty miles northwest to Attu. This rugged island, indented by fjords and almost forty miles long, was home to about thirty Aleut men and their families.[17] It was also a summer home to nesting murres, tufted puffins, arctic terns, black-legged kittiwakes, red-faced cormorants, eider ducks, and Aleutian Canada geese. Rock ptarmigan lived on the slopes of 3,000-foot Mt. Attu year-round.[18] The surrounding waters were rich with sea lions, seals, and sea otters.

Russian explorers dealing with potentially hostile indigenous people had developed a custom of taking hostages. During the year and a half this first expedition spent on Attu, the trappers sequestered in their camp a few young Aleuts to "teach them Russian." At the same time, the Russians were learning from the Aleuts. They adopted Aleut outerwear, diet, and housing, but they could not match the Aleuts' skill in hunting on the sea. Most of the sea mammal pelts they accumulated were acquired by the Aleuts. In the end, the hostages were returned and the Aleuts received trade goods. But on Attu, as on Agattu, there had been a bloody incident, and mistrust between the two peoples remained.

For Tolstykh's fox transplants to be successful, he needed good working relations with the Attu Aleuts. For the first three or four years, the foxes had to be left to propagate undisturbed. A well-fed vixen could raise a litter of four to eight pups per year. By one year of age the pups would be full grown and able to raise their own litters. When the population burgeoned, Tolstykh would supply the Aleuts with traps to capture foxes at midwinter when the fur was prime, carefully leaving behind a breeding population.

Fortunately, Tolstykh had diplomatic talent. He presented Tunulgasen, the chief of Attu and the other Near Islands, with a copper kettle and a full set of Russian clothing, including boots and a fur coat. He offered typical trade goods—needles, hatchets, knives, wool, and beads—not just in payment for furs but also for permission to stay on the island.[19] Tolstykh and his crew spent two winters on Attu. Two Siberians from the ship were killed by Aleuts in a brief conflict, but no revenge against the village was exacted. Tolstykh and Tunulgasen parted on good terms.[20] The foxes, omnivorous hunters and scavengers, settled into their new environment, digging burrows for homes. In the summer, they fed enthusiastically on nesting birds and eggs, even storing some eggs for fall. The foxes were agile climbers; only the highest cliff nests were safe from these vulpine hunters. In the winter, foxes subsisted sparely on the remains of sea mammals butchered by the Aleuts and on food from the ice-free beaches: sea urchins, sea squirts, wild parsnip, grass, and the occasional dead fish or beached whale.[21]

In 1757, Tolstykh returned to Attu, this time equipped with Siberian-style spring traps constructed from whale sinew and wood that was studded with iron teeth. The Aleuts set the traps, unbaited, in clever locations.[22] After the first trapping season, islanders began dragging seal carcasses onto the beaches as supplementary food for the breeding foxes.[23] Tolstykh based his work crews on the island for a year. When he returned to Kamchatka, he unloaded 5,360 sea otter pelts and 1,190 blue fox pelts.[24] However, Attu was not Tolstykh's private reserve. Although he spent two subsequent winters there, other ships' crews also spent seasons on the island catching sea mammals and trapping foxes.

Following the success on Attu, Russian fur traders transplanted blue foxes from Attu to other Aleutian islands, including Atka, Amilia, and Kiska. These arctic foxes were not only being moved east but also south toward the center of the curving Aleutian chain, the most southerly point in Alaska.

The climate of these central islands—located at about the same latitude as London—soaked the foxes' fur and taught them to hunt a wider range of beach prey and to aggressively scavenge food left intentionally or accidentally by humans. On islands with large summer bird populations, the arctic foxes not only adapted but thrived.

Native Silver Fox Transplants

In the meantime, in 1759, Russian entrepreneurs had pushed even farther east and discovered the Fox Islands near the Alaska Peninsula. This group of islands is named not for the arctic fox but for the larger, longer-haired red fox, *Vulpes vulpes*, native to the Fox Islands as well as the Alaska mainland and most of North America. The genetics of hair color in the red fox family are complex, but it is possible to breed a consistent strain.[25] The three usual color patterns of this fox are described as red, cross (mostly red with a black stripe down the back and across the shoulders), and silver. The Russians and other

Silver-black fox (*Vulpes* genus) with long legs, pointed ears, and white tail tip, photographed in summer.
Alaska State Library, Talmage Family Photo Collection, P345–241.

early writers often referred to this latter group as black foxes or silver-black foxes because some of their black guard hairs had a central white band that gave the fur a silvery sheen that could not be reproduced by dyes.

Silver foxes were both the most valuable and the rarest of the red fox tribe, making up from 2 to 17 percent of regional fox populations.[26] On the Fox Islands and nearby Alaska Peninsula, the proportion of silver foxes was at the top of this range. When marketed in Kiakhta, these silver fox pelts sold for twenty to thirty rubles, in contrast to red fox pelts, which sold for a ruble or less.[27] Initially, Alaska silver foxes were simply trapped and pelted, although in 1764, trader-explorer Stepan Glotov brought a breeding pair from Alaska to Siberia with the intent of establishing a population closer to home. The outcome of this attempt is unknown, but on the same trip Glotov also presented the Tsarina, Catherine the Great, with a number of fine silver fox pelts. She showed her pleasure with the gift by awarding medals to Glotov and his co-investors, and she canceled a debt owed to the government for outfitting their vessel.[28] Glotov's present also had its intended effect on St.

Blue arctic fox (*Alopex* genus) with short legs, compact body, and rounded ears, photographed in summer.
Alaska State Library, Talmage Family Photo Collection, P345-233.

Petersburg fashion: The value of silver fox pelts increased.²⁹ This economic ripple returned to Alaska, and within a few years Russians would be transplanting native silver foxes onto various "unused" Alaska islands.

The Pribilof Islands: A Blue Fox Enclave

For many years, however, silver fox pelts were too expensive and too scarce to supply much of the demand for elegant winter clothing. Blue fox continued to dominate the northern European luxury trade. The dark fur was used to line fashionable wide-sleeved capes. It trimmed dresses, coats, and—in those drafty days—even petticoats. Stylish women wrapped narrow fur boas around their necks and wore dark fox neckbands to emphasize the white expanse above their décolletage. They carried capacious fur muffs, large enough for a lapdog as well as for milady's forearms.

There seemed to be no end to demand and every reason to plant blue fox colonies on each "empty" but fox-habitable island of the western Aleutians. There was no shortage of suitable islands, but the Russian entrepreneurs were stymied by the small number of blue foxes available for transplant. By the 1760s, blue foxes were no longer abundant in the Commander Islands.³⁰ Nearly all breeding pairs had to be captured from Attu, where the colony was well established but relatively small. Finally, in 1786, the problem of finding sufficient breeders was solved. Incidental to the fur seal trade, a new and unsuspected source of blue foxes was discovered.

By this time, Russian seamen had hunted sea otters into deep decline and had shifted their efforts to a different pelagic prey—the northern fur seal, whose large, easily dyed pelt had become popular for coats. The swimming seals were not hard to kill, but they were hard to bring aboard. Some estimates suggest that as few as one in seven animals killed was actually hauled onto a boat.³¹ Russians regretted their loss of the now-depleted fur seal rookeries of the Commander Islands where the seals had been easy to corner on the rocks and kill with clubs. On land the hunters could even identify and choose to spare females. Nevertheless, greed had almost demolished the Commander Islands herd.

In Alaska, fur seals had been seen only in the water, never on land. They passed north among the islands of the Aleutian chain each spring and south each fall. The Russians reasoned that the herd's movement was a migration driven by a winter journey to some rich feeding area and a summer return to mating and breeding grounds. They guessed correctly that mainland or island rookeries lay to the north and surmised incorrectly that rocky islands on which seals could haul out during the winter must lie to the south. The exact course of the spread-out, milling herd of nearly two million animals was unclear. For two decades, captains searched for rookeries and haul-outs. For reasons of weather, more searchers ventured south into the Pacific than north into the foggy, storm-prone Bering Sea.[32]

However, one of these searchers, Gavriil Pribylov, listened attentively to an Aleut story passed through many generations. A hunter, Igadik, had been driven north from Unimak Island by a winter storm and fetched up on an island he called Amiq. There he spent the spring and summer hunting fur seals in rookeries.[33] Pribylov, son of one of Bering's sailors, was a navigator for the Lebedev-Lastochkin Company in charge of a sturdy, shallow-draft galiot, *Sv. Georgii Pobedonosets (St. George the Victorious)*. He had a reputation as an unusually skilled mariner,[34] and he had enough experience in the region to not be dismissive of Aleut "myths." But his search for Amiq required persistence.

In 1786, on his third summer of searching, Pribylov found himself about 230 miles northeast of Unimak, and within earshot of his goal. Through fog so thick that he could barely see the length of his ship came bawling, grunting, and splashing from a nearby rookery. One story claims that the land only became tangible when his bowsprit blindly bumped against the base of a 350-foot cliff. Another account has the fog lifting to reveal a twelve-mile-long volcanic island with luxuriant tundra and rocky headlands, an island that during the summer hosted a million nesting birds (more than 200 species) and more than half a million fur seals. The discovery on June 25, 1786, was worth at least as much as the imaginary value of the spurious DeGama Land for which Andrean Tolstykh had fruitlessly searched. Pribylov called the island St. George, after his ship.

On shore, the exploring party found a bonus. The island was year-round home to arctic foxes, most of them blue. Like the Commander Islands, the island of St. George was connected by ice to the mainland 300 miles away only during rare winters. During one or more of these unusual winters, arctic foxes, wandering across the ice to scavenge the remains of polar bear kills, had settled on the island. Some of the fox ancestors carried the dominant blue gene for coat color, no great disadvantage for island hunting. Inbreeding had caused their blue descendents to multiply.

St. George Island did not have a protected harbor, so Pribylov sailed away for the winter, leaving behind a hunting crew of about forty Aleuts and Russians. The following spring, members of this crew climbed a small peak in the hope of spying their ship on its return. There was no ship in view, but when they looked north, they sighted an island about 40 miles away. This brother island, soon to be named St. Paul, had no human inhabitants. However, like St. George, it supported a bustling summer population of breeding fur seals and nesting birds. And in winter, most of its arctic foxes showed up dark against the snow. The Pribilof Islands, as they came to be called, were the birthplace of millions of birds and four-fifths of all northern fur seals. They also had the highest concentration of New World blue foxes west of Greenland.[35]

Russian-American Company Fox Farms

The crew from Pribylov's discovery voyage spent two full years on the islands.[36] But in later years, other fur hunters, some of whom almost literally sailed in Pribylov's wake to find this mother lode of furs, came only for spring and summer. The Aleut sealing crews they brought to the islands were mainly seasonal workers. Their take of blue fox pelts, prime only in winter, was initially low. But in 1799, the Russian-American Company took control of the Alaska fur trade. This quasigovernmental monopoly not only attended to fur harvests, but also dispensed justice, bolstered the church, and managed people within the colony. With promises, bribes, and coercion, the company transplanted Aleuts from Unalaska and Atka to establish villages on St. George and St. Paul. The Aleuts dug in, creating year-round semi-subterranean homes. Managers' quarters, warehouses, and eventually two churches were built from beach logs and imported lumber. Foxes on both islands, feeding on heaped-up fur seal carcasses, raised large litters and the population burgeoned. Winter trapping began in earnest.

In the first two decades of the nineteenth century, while the Aleut communities were getting established, the number of blue foxes pelted in the Pribilofs averaged about 600 per year.[37] During the remainder of the Russian period, the yearly average ranged between 1,900 and 3,000.[38] In addition to harvesting pelts, the Russian-American Company captured live foxes for transplant. The Pribilofs provided a wealth of breeding stock for various Aleutian Islands. In 1819, the company's St. Petersburg headquarters ordered 900 pairs of blue foxes to be released on "lesser Aleutian Islands." The following year, another 200 pairs were to be transferred to each of the Rat Islands (Kiska, Amchitka, Segula, and Semisopochnoi).[39] It is not clear whether these orders were fulfilled in the quantities indicated, but records suggest that over the years Russians shipped more than a thousand live foxes from the Pribilofs to other islands.

The Russian-American Company did not limit its propagation efforts to blue foxes. In 1842, the company also directed that silver foxes from the Alaska mainland be transplanted to "the principal islands of the Unalaska and Atka Districts" (the Fox, Andreanof, and Four Mountain Island clusters). These orders were followed by instructions to send live ground squirrels from Kodiak Island to islands with new fox populations.[40] This attempt to provide a nonmigratory supply of food for the foxes was not altogether successful. The ground squirrels adapted well but disappeared into hibernation during the lean winter months and competed with the foxes for bird eggs in the summer.

Although the Russian-American Company never referred to the islands as "farms," translators indicate that they did refer to the foxes as a "crop" and used "harvest" to describe the trapping, killing, and skinning of the animals. Certainly, rudimentary elements of farming were in place. The foxes were fenced by the sea. Natural foods were supplemented by seal carcasses or ground squirrels. The foxes, although not domesticated, were habituated to people to the point of being a nuisance, boldly seeking unattended edibles in settlements. Islanders also managed their crop by avoiding over-trapping, which would threaten breeders. They killed raptors—eagles and ravens—that threatened pups when they first emerged from the dens. And, in their own interests, they defended grown foxes from poachers.

The Russian-American Company also practiced conservation on these lucrative islands. Kiril Khlebnikov, company manager in Sitka from 1818 to 1832, wrote, "Prudent moderation . . . is the most useful method for secure profit. . . . Should the hunting of foxes . . . be increased, it is almost certain that . . . the catch would double. But what will remain for future prospects?"[41] When the Russian-American Company perceived that fox populations were low, they imposed one- to two-year moratoriums on fox trapping. When the company noted an increase in undesirable white and dingy-gray foxes on the western islands, they established a year-round bounty on white animals equal to the value of a prime blue fox pelt.[42]

The early independent entrepreneurs such as Glotov and Tolstykh died before the Russian-American Company took over and added these refinements to managing the pelt harvest. Stepan Glotov, a crew member on a near-disastrous expedition to chart the Alaska Peninsula, died (probably of scurvy) on Unimak Island in May 1796. Gavriil Pribylov, who continued to captain ships sailing as far south as the Queen Charlotte Islands, also died in 1796 of an unnamed chronic illness in Kodiak.[43] Andrean Tolstykh had died thirty years earlier. During his two decades in the Aleutian fur trade, Tolstykh's fortunes had fluctuated. He progressed from skipper, to captain-investor, to ship owner and, after a shipwreck, back to hired skipper.[44]

In 1766, perhaps because he had found rough charts made by other Russian skippers of portions of the Aleutian chain to be trustworthy, Tolstykh decided to make another search for the DeGama Land that appeared on the 1733 chart of the North Pacific. When he returned from this second unsuccessful search for the fabled island, Tolstykh anchored off the Kamchatka coast during a storm. The anchor failed, the ship broke up on near shore rocks, and Tolstykh did not survive to reach the beach.[45] Just a few months before his death, Catherine the Great, who had previously remitted his 10 percent fur tax for locating the Andreanof Islands, honored him a second time in an edict.[46] Later, historians also praised Tolstykh. Hector Chevigny found him unusual in his fairness and warm relations with the Aleuts and regretted that "his methods were adopted by few." Vasilii Berkh called him "the best seafarer of his time." Richard Pierce quotes A. I. Alekseev's opinion that had Tolstykh survived, he would have become Russian-America's premier fur trader.[47]

Aleut Warning: Foxes Destroying Bird Colonies

Neither Catherine the Great nor the historians who honored Tolstykh recognized the fox transplants as an important legacy. Island fox farming, which continued after the sale of Alaska, was to have far-reaching consequences for the natural history of the Aleutians. The Aleuts were the first to sound the alarm, complaining about the decrease in nesting birds on islands occupied by foxes. A warning from Atka was recorded by the navigator Ivan Vasil'ev in 1811: "The foxes drive away the birds, which formerly were very numerous and served as a source of feathers for clothing. Nowadays, to get birds they [the Atka Aleuts] must travel to other islands."[48] To make clothing, Attu Aleuts in this period turned more and more to the use of fish skins to replace hard-to-obtain bird skins.[49]

These early warnings were recorded but forgotten. Birds had no commercial value. In fewer than sixty years the Russians would sell Alaska, leaving descendents, churches, schools, and foxes behind. Fragmented records show that at the time of Alaska's sale in 1867, foxes were being raised and harvested on at least seven Aleutian islands in addition to the Pribilofs.[50] After the withdrawal of the Russians, the untrapped foxes would exhaust bird colonies and die out on some of these islands. On other islands they would reach an equilibrium that made them appear indigenous. As the paperwork of Russian America gathered dust, few of the new English-speaking owners of Alaska learned the history of the Aleutian foxes or about their effect on bird populations. And the little that was learned was quickly forgotten.

In 1870, three years after the American purchase, William Dall, head scientist for the Western Union Telegraph Expedition, wrote a book, *Alaska and Its Resources*, to convey "present knowledge of Alaska." He stated that blue foxes were introduced to the [Aleutian] Islands by the Russian-American Company. "They [the foxes] have exterminated all the small animals, if any existed . . . and feed on sea-birds or the carcasses of seal abandoned by the natives."[51] But even this imperfect knowledge faded rapidly. Some thirty years later, when the results of the Harriman scientific expedition to Alaska were published, Dall was one of multiple authors, but he did not write the section on fox farming. The author of that chapter was M. L. Washburn, a Kodiak entrepreneur who made no mention of Russians in his explanation of the origins of Alaska's fur farm industry. His version, described in the next chapter, would be widely accepted in early American Alaska.

CHAPTER TWO

The Pioneer Period 1867–1909

Americans Reinvent Alaska Fur Farming

Roughly ten years after the Russians left Alaska, an energetic 22-year-old from Vermont arrived in Kodiak. M. L. Washburn knew little about Alaska or its Russian history and nothing at all about farming foxes.[1] Ready for adventure, Washburn had made his way from the East Coast to the headquarters of the Alaska Commercial Company in San Francisco, where he was hired as clerk and bookkeeper for its Kodiak store. From his new post, he surveyed the surrounding resources with entrepreneurial curiosity.

Washburn possessed energy and business acumen. Within two years— due in part to his boss being shot by a demented man before Washburn's eyes at the breakfast table—he was made manager of the Kodiak district for Alaska Commercial Company. Within five years, he and several partners had established Alaska's first American fox farming company. He began managing one of the company's blue fox farms near Kodiak, where he continued to work for the Alaska Commercial Company. In 1899, approximately twenty years after arriving in Kodiak, the convivial Washburn was recruited to travel from Cordova on Prince William Sound to Kodiak with the Harriman Alaska Expedition as their fur farm expert.

The Harriman Expedition, traveling the coast from Seattle to the Bering Sea in the steamer *George W. Elder*, made headlines in the States and a sensation in Alaska. The *Elder* had been chartered for a summer trip to Alaska by railway magnate Edward Harriman after he was advised by his doctor to take a vacation. The *George W. Elder* was a luxurious vessel equipped to carry 126 passengers and crew. Harriman's family was small, but instead of filling the ship with friends, he converted the vacation into an unusual scientific expedition, adding at his own expense two dozen eminent scientists and artists to the passenger list. The staff on board included stenographers, hunters, a taxidermist, and a chaplain. The library held more than 500 Alaska books, and food was prepared by a skilled chef. The Harriman scientists and the reports that they would write were destined to influence decades of government decisions relating to the land, game animals, and fur farms in Alaska.

M. L. Washburn wrote the final chapter in the Harriman Alaska Expedition's two-volume report. His chapter, "Fox Farming in Alaska," would prove to be a popular source for future writers, and his description of how fur farming began in Alaska would be widely repeated.[2]

> Fifteen years ago a few men in western Alaska, realizing that fur-bearing animals were doomed [by over-trapping], decided to try the experiment of propagating . . . the blue fox . . . by placing a small number on protected islands and caring for them as the stockman cares for his herd of cattle. . . . About twenty foxes were taken from . . . the Pribilof group, and placed on North Semidi, one of the hundreds of unoccupied islands of Alaska.[3]

Washburn was one of the "men in western Alaska" whom he credits with first farming blue foxes. He and three partners founded the Semidi Fox Propagation Company in the years between 1881 and 1885 (sources vary). After their Pribilof blue foxes were well established on North Semidi Island fifty miles east of the Alaska Peninsula, they transplanted Semidi offspring onto five additional islands south and east of the peninsula. Washburn's partners—W. B. "Preach" Taylor, Thomas F. Morgan, and James Redpath—had each worked for the government or the Alaska Commercial Company in the Pribilofs, where the U.S. Treasury Department managed the fur seal and blue fox harvests.[4]

Government Fox Sales and Fur Farm Leases

The first three harvests of Pribilof fur seals after the purchase of Alaska had been a resource disaster—a capitalist free-for-all by competing companies that took a quarter of a million fur seals without regard to the risk of extinction. The U.S. government became alarmed by the slaughter. In 1870, Congress designated the Pribilof Islands a reservation under the supervision of the Treasury Department.[5] Furs were the chief resource the United States had expected to gain by buying Alaska, and the Pribilof rookeries were considered particularly valuable.[6] Treasury agents were charged with regulating the take of fur seals and foxes on the reservation to produce maximum sustained yield for the coffers of the nation. To make this oversight feasible, only one company was allowed onto the islands to harvest furs. Congress issued the first twenty-year contract to the Alaska Commercial Company, which had acquired ships, buildings, and other property from the Russian-American Company.

The contract required the company to provide frame houses for the Aleut families whose men harvested fur seals, trapped foxes, and prepared the skins for shipping. These inhabitants, numbering about two hundred, were now two or three generations removed from their origins in Atka and Unalaska. They had come to consider themselves residents of the Pribilofs, not immigrants from some other island. With the establishment of the reservation, they became wards of the U.S. government. The government supplied a doctor and one teacher for each island. The Alaska Commercial Company built the schools. Except at the islands' Russian Orthodox churches, English replaced Russian as the second language of the people. Workers were paid small amounts of cash for their labor, but families had to rely on sometimes-inadequate, government-issued food to supplement subsistence fare. The cost of labor for Pribilof fur operations was kept low for many decades. Books such as *A Century of Servitude* (1980) by Dorothy Jones and *Slaves of the Harvest* (1983) by Barbara Torrey describe the plight of Pribilof Aleuts during the century that U.S. government agencies supervised the islands, their animals and their people.

When the treasury agents took over the Pribilofs, their first action in trying to rebuild the fur seal herd was to cut drastically the allowable harvest. In some years, the harvest was limited to as few as 7,500 animals.[7] As for blue foxes, the agents relied on the Aleuts to estimate the population and determine appropriate trapping numbers. At first, the fox harvest was similar to Russian years—over a thousand pelts per year. In the early 1880s, when M. L. Washburn and his partners sought to buy Pribilof foxes for transplant, agents had no difficulty granting the request. They had sufficient foxes and wanted to support any new industry within the District of Alaska. They supplied the Semidi Company with breeders at a bargain price—$10 per pair plus shipping.[8] Fifteen years later, the Semidi Company would sell breeders to other farmers for $100 to $250 per pair.[9]

The treasury agents on the Pribilofs were not troubled by selling live foxes from a government reservation to private citizens, but they were concerned about the islands that the Semidi Company (and perhaps others) would need if fox farming caught on. The Treasury Department was not only responsible for the Pribilof reservation, it was also in charge of nearby land in the Aleutians and along the coast.

A dilemma for pioneer fox farmers was how to obtain access to islands for farming. All land in Alaska, with the exception of a few mining claims and townsites, belonged to the federal government. The Homestead Act had not yet reached Alaska; and when it did in 1903, it would retain a clause disallowing homestead claims on land "occupied for the propagation of foxes."[10] There was also no mechanism for outright purchase of government land. Some farmers simply "squatted," declaring a property right by actively using an island.

The problem of access to land for fur farmers in the eastern Aleutians and nearby coastal areas was solved in 1882 when the Treasury Department began offering five-year fur farm leases. A farmer was instructed to identify a desirable, uninhabited island. If no one else had applied for it, treasury agents would lease the island—regardless of size—for $100 per year. Paid-up leases could be renewed as long as the island was actively farmed. From the government's point of view, official leases would minimize squatters' squabbles, and if an island ceased to be used, it could be made available to another entrepreneur.[11] Leasing also appealed to many farmers. Leaseholders gained a legal right to bar poachers and other uninvited visitors to their islands. Moreover, a farmer did not have to make a large initial purchase payment or a long-term commitment. The Aleuts who used these "uninhabited" islands seasonally or intermittently for subsistence were not consulted.[12]

Before long, notices were posted near the beach on leased islands and in surrounding towns at the post office or village store. The signs published the right of leasees to keep others off their islands.

> The following islands in Alaska have been leased by the Government of the United States to the parties named. . . . All persons are warned not

to land upon any of these islands, except upon written permission given by the respective keepers of each island . . . no dogs or firearms . . . are allowed. . . . A reward of ONE HUNDRED DOLLARS has been offered for evidence to secure conviction of . . . trespassing.

These warnings were produced by the Fox Breeders Protective Association of Alaska.[13] Their list of leased islands was limited to a dozen, four of which were registered to Semidi Company partners. But many fox farmers were using islands without taking out a lease. In 1897, a government official traveled from the Alaska Peninsula to Prince William Sound and listed twenty-four active fur farms. He noted that fully half the farmers had declined to pay the $100-per-year rent, which he considered "nominal," and were using government land without permission.[14]

The Semidi Fox Propagation Company on Long Island

As the most prominent fur farmers in Alaska at the end of the century, M. L. Washburn and his partners had properly leased islands. Washburn was personally in charge of the Semidi Company's blue fox farm on Long Island near Kodiak.[15] About a year before the company formed, Washburn had made a failed effort to raise foxes. He tried to nurture a captured litter of silver foxes in a pen, but none survived. He theorized that a kindly keeper had overfed them, and he did not repeat the penned silver fox experiment.[16] Raising blue foxes on an island proved easier and more fruitful. By 1899, when Washburn wrote the "Fox Farming in Alaska" chapter, he had gained considerable expertise.

The Long Island farm, one of six operated by the company, was seven miles as the raven flies from Kodiak. A keeper and two assistants managed day-to-day operations. Washburn made only periodic visits because the boat route was circuitous and time consuming. Even the modern "naptha launch"[17] belonging to the Harriman Expedition required an hour to reach the island from Kodiak.

Washburn described Long Island as "perhaps the ideal blue fox ranch and may serve as a type for all." (In fact, it did serve as a prototype for *Rocking Moon*, a 1925 novel and 1926 silent movie, whose heroine was Sasha Larianoff, a beautiful Alaska fox farmer.)[18] The foxes, a few sheep, and about forty cattle ranged freely through the grass, berry bushes, and scattered spruces on the hilly, three-square-mile island. On the island and in the ocean natural fox foods abounded: cod, halibut, salmon, flounder, clams, mussels, berries, and nesting seabirds. At the north end where the beach grass grew tall there was a natural harbor protected by a rocky islet at its entrance. The harbor was bordered by frame company buildings—a keeper's house, a skinning shed, fish-drying racks, and a warehouse for equipment and the sacks of cornmeal that lent bulk to cooked wild foods. The farm was picturesque and productive.

In keeping with the times, Washburn's account of fox farming is infused with business confidence and enthusiasm. Alaska was America's newest frontier. Fur farming was her newest business. Fur sales in Seattle were booming. Prospectors rushing to the Klondike were buying up fur mitts, hats, and coats before leaving the States. Rumor said that all the trappers up north had quit their traplines to work on mining claims.[19] Washburn and his fellow entrepreneurs were in the right place at the right time. In Seattle and in London, where most Alaska furs were sold, prices were high for dark fox pelts: $5 to $50 for blue fox and $100 to $200 for silver fox. Rare silver fox pelts, fully black except for tail tip, were worth over $2,000.[20]

London, thanks to the Hudson's Bay Company, had become the busiest fur market in the world. Its auctioneers, C. M. Lampson & Company, were adept at contracting with distant traders and fur farmers. The marketing system worked on trust. Sellers shipped lightweight boxes of raw furs to Lampson. The company numbered and inspected each pelt. The pelt was then cleaned with sawdust, beaten, combed, and turned inside and out to soften the skin. Each numbered pelt was grouped with others of similar value from the same seller to make up one lot.

Before the sale, qualified buyers came to a large room where the lots were displayed. Wearing white coats to protect their clothing from grease, buyers recorded careful notes. At the sale several days later, no skins were displayed. Quiet, almost silent bidding determined the price of each lot. Although sellers could request a 60 percent advance on the probable value, most Alaskans simply waited until the check (with deductions for Lampson's charges) arrived about a month after the sale.[21]

A government bulletin of this period validates Washburn's confidence in the enduring value of furs: "In warmth, beauty, and durability no manufactured

The Pioneer Period 1867–1909

Fairbanks woman poses in stylish fox stole and ostrich-feather hat, ca. 1912.
Alaska State Library, James Wickersham Collection, P277-11-38.

fabrics excel them."²² The growing middle and upper classes on both sides of the Atlantic agreed. Gentlemen wore beaver hats, the ladies sported seal toques, and almost everyone—even of modest means—owned a winter coat with a broad fur collar. In the evening, wealthy women wrapped themselves in dark fur stoles with dangling tails or donned wide-shouldered, narrow-waisted fur jackets. Fox muffs, although smaller than those during the Russian colonial period, remained popular. Demand for utilitarian and luxury furs was high. But wild furs were becoming scarce due to loss of habitat and over-trapping. Washburn and other farmers saw that fur farms had exciting potential for replenishing the market. Some nascent conservationists predicted that fur farmers would eventually replace trappers, and wild furbearers would again roam undisturbed in the wilderness.

Although blue foxes were less valuable than their silver cousins, the Semidi Company and other Alaska blue fox farmers at the end of the nineteenth century felt their position was advantageous. The climate produced thick fur. Enclosed by the sea, Alaska's islands provided abundant beach edibles and summer birds to supplement farm rations. Locals who understood the climate were readily hired to live on islands as keepers. After a farm was established, the owner could sell furs to the market and live foxes to newcomers who had just leased one of the "hundreds of now useless islands" from the government.²³

James Judge and the Pribilof Fox Farm

By 1899, Washburn and other private fox farmers were also profiting from lessons learned by the U.S. government on the Pribilof Islands. In the early 1890s, treasury agents on the islands were alarmed by an unaccountable decline in the blue fox population. Although they suspended sales of breeders and stopped shipping live foxes off the islands, pelt numbers fell from historic levels of over 1,000 to a low of barely 300.²⁴ Baffled, the agents speculated that foxes were walking away on drifting ice or that the Aleuts responsible for estimating fox numbers had miscalculated in previous years, causing the animals to be over-harvested.

Then in 1894, James Judge, a 28-year-old assistant treasury agent, was assigned to St. George Island. He arrived with his wife, Helen, who would eventually become an expert in Pribilof ornithology. Both the new arrivals were observant naturalists. In an effort to learn more about blue foxes, they began to raise a tiny pup in their house.[25] After just a few months on the job, Judge felt he understood why the island foxes were becoming scarce. In 1892, an international treaty had placed a long-term cap on Pribilof fur seal hunts. One result was that fewer seal carcasses littered the beaches.[26] The foxes were suffering from a shortage of food. They gorged on eggs, birds, and carcass meat during the summer. In the fall when the birds flew south and the meat ran out, they turned to beach edibles. But when drift ice and freezing spray coated the beaches, there was little to carry them through the lean winter. Many died.

Judge took action. The following summer he and Aleut workers cut up and salted some of the fur seal carcasses for use when the days grew short. They stored the meat in tunnels dug into a cold hillside. But foxes cannot eat salted meat. Judge's innovation was fitting each tunnel with a covered hole in the roof through which fresh water could be poured when the food was

Small feed house/trap, similar to those developed by James Judge. The ramp leads to the fox entrance on the upper floor. The front screen window opens for placement of food. The spring-loaded trapdoor inside drops foxes into the lower holding room at pelting time.
Photo courtesy of Annie Nilsson

needed. Workers ran water siphoned from a stream through the meat and out through the tunnel entrance for two weeks, then they distributed the "freshened" meat to covered feeding stations on the island that protected the fox food from ravens and gulls.

Judge made other improvements that were noted by private farmers. He replaced the deadfall and leg-hold traps that killed nonselectively with a system of feed-house traps that allowed the best stock to be saved for breeding. The feed houses had a ramp to a large upper chamber that was entered through a door tied open with rope. In the fall, the chamber was supplied with meat, which the nocturnal foxes were quick to locate. When winter came and coats were thick, an Aleut trapper would wait quietly in the darkness with the rope to the door in his hand. After several foxes were feeding, he would lower the door. An assistant then herded the animals into a large "retaining room." When twenty or so animals had been captured, they were released a few at a time into a corral where a heavily gloved worker with a U-shaped crook trapped each fox and lifted it by the nape of the neck so that a recordkeeper could note the animal's sex, age (based on a cautious examination of the teeth) and weight (using a suspension scale). The data were used to identify the best animals for breeding.

Each season 500 foxes were selected as breeders on the basis of hefty weight, dark glossy fur, and prime age. They were "branded" by clipping off a ring of tail hair—near the tip for males and near the base for females. Marked foxes were released. If they reappeared in the trap, they were simply shooed out of the corral. All white foxes were culled, and blue foxes not selected for breeding were killed for pelts. James Judge remained in the Pribilofs for nineteen years and published at least one article about his methods. During his tenure, white foxes on St. George declined to 1 percent of the population, and blue fox pelt sales gradually rebounded to 1,200 per year.[27]

In his fur farming chapter in the Harriman Expedition report, M. L. Washburn credits methods developed on St. George when he describes the Semidi Company's thriving Long Island farm. The original sixty foxes transplanted five years previously had increased to a herd estimated at 800 to 1,000. These relatively tame animals gathered from all parts of the island for their once-daily feeding of cooked fish and corn meal prepared by the keeper and his assistants. In midwinter, when the fur became prime—thick, blue-gray, and glossy—the foxes were caught in box traps. (In future years,

Long Island would convert to elevated feed houses with spring trapdoors in the floor. When the trapdoors were unlocked during the pelting season, animals walking over the door dropped into a holding cage and the door sprang shut ready to trap again.) As on St. George, Washburn's keepers kept careful records and marked the best individuals as breeders by clipping a patch of tail hair. Foxes were killed for their pelts only if they were not suitable as breeders because healthy, live animals commanded a higher price than pelts. In 1903, when fox prices took a substantial dip, the average Long Island pelt sold for only $8.70, whereas a healthy breeder brought at least ten times that amount.[28] By 1900, the Semidi Company had already sold enough breeders to stock seven islands in addition to their own.[29]

Thomas Vesey Smith on Middleton Island

Selling breed stock was a growing business for all established fox farmers. By the end of the nineteenth century, blue fox farming was expanding from areas near the Alaska Peninsula westward into the Aleutians, where Samuel Applegate placed Pribilof foxes on Samalga Island in 1897. Fox farming also spread eastward along coastal Alaska. Washburn estimated that by 1899,

Blue foxes forage on beach of island fur farm.
Lulu Fairbanks Collection, UAF 1968-0069-00097, Archives, University of Alaska Fairbanks.

thirty Alaska islands had active farms. In Kachemak Bay, at the tip of the Kenai Peninsula, blue foxes were planted on Hesketh and Yukon Islands in 1900. In Prince William Sound in 1894, a Swedish immigrant, Fred Liljegren, paid $1,800 to bring twenty-four blue foxes from Greenland to Storey Island. By 1903, he and his partners not only sold breed stock to new farms in the area but also harvested 165 pelts—more than any other privately owned Alaska farm.[30] Blue foxes were also transplanted along the coastal curve and into the dense rainforest of southeast Alaska, a climate in which arctic foxes can survive only with human care. In 1901, James York planted thirty pairs of blue foxes on Sumdum Island near Juneau, and Mrs. George Scove started a blue fox farm on Patterson Island near Ketchikan.[31]

Journalists writing about these early island fur farms often snared their readers by emphasizing the solitary nature of the farmer's life. One told of a farmer who not only "sought distant exile" but also "reinforced his seclusion by the angry barriers of the tempestuous northern sea."[32] Another was even more dramatic: The fox farmers of Alaska are "the most exclusive people on earth. . . . Theirs is a life of loneliness, ostracism, exile, desolation."[33] The

Farmer at island near Petersburg feeding foxes from a boat.
Lulu Fairbanks Collection, UAF 1968-0069-0098, Archives, University of Alaska Fairbanks.

Alaska fur farmers, however, did not see themselves as hermits escaping society; instead, they considered themselves pioneers at the forefront of an industry producing a valued commodity in a new way. They were counting on a personal reward for their vision and hard work. It was incidental that they were not afraid of solitude.

The most isolated turn-of-the-century farm was grassy Middleton Island in the Gulf of Alaska about 130 miles south of Valdez. Middleton, seven miles long and a mile and a half wide, had no harbor.[34] It was a summer nesting home for gulls, auklets, murres, kittiwakes, puffins, guillemots, and cormorants. The first farmer, P. R. Temple, started in 1890 with silver foxes. In 1903, he sold to Thomas Vesey Smith. Smith had arrived in Alaska from Maine, where his skill as a mariner on sailing ships was no longer in demand. He landed in Valdez with the intention of trekking to the gold fields of Interior Alaska over the Valdez Glacier. But when he fully grasped the distance and geography of his proposed trek, he found the proffered Middleton fox farm, accessible by sea, a more attractive proposition. The previous owner had traveled to and from the island by hiring a motor launch, but Smith was a seaman. For transportation he bought a shallow-draft twenty-eight-foot sailboat that could be beached for the winter.

Guided by a chart sketched by the seller, Smith and a partner sailed to Middleton Island, stopping at Storey Island to buy blue foxes from Liljegren. To Smith's dismay, the Greenlandic blue foxes soon eliminated the more valuable silver foxes, and before long his partner succumbed to insanity. But Smith persisted alone with the foxes as close companions (including a litter of seven that grew up in his house). His herd of blue foxes increased to 200. He hauled a homemade feed cart to fox burrows on distant parts of the island. He caught fish, shot seals, and raised ten acres of potatoes for himself and the foxes. Whenever a storm roiled the beach and revealed black sand, he panned it for a small return of gold.[35]

In May, while the foxes subsisted on bird eggs, Smith returned to Valdez to sell his furs, buy provisions, and pick up his mail (including a year's worth of *Philadelphia Ledger* newspapers that he would read on his island one year late to the day). During his 1908 trip to town, poachers, believed to be Japanese fishermen, not only killed many foxes but also took valuables and set fire to his cabin. Nevertheless, Smith persisted for eight years and left Alaska with enough money to buy a ranch in the state of Washington.[36]

The Fur Farms of Alaska

Silver Foxes in Pens

Although Alaska's infant fur farm industry was dominated by blue fox farmers, a few Alaskans chose to raise silver foxes. The silver foxes were more valuable than blues but harder to tame. They had a reputation for being aggressive, nervous, and hard to entice into traps. All vixens will kill their pups if they feel threatened, and silver foxes were easily unnerved. During whelping, strangers on the farm, the sight of a stray dog, or the sound of thunder, hammering, or a gunshot could result in disastrous losses. And serious problems arose when farmers tried to raise blue and silver foxes together, although in contrast to Smith's experience with the Greenlandic blue foxes, the silver foxes commonly eliminated the blues by killing them outright or by out-competing them for scarce winter food.

Unlike blue fox farmers, early silver fox farmers in Alaska did not have a convenient supply of partially domesticated breed stock. At first, farmers paid handsome prices to trappers for live animals. Prices were so high that for some trappers, catching a silver fox was compared to a winning lottery ticket.[37] This, of course, led to abuses—out-of-season trapping, digging up dens, and even dyeing red foxes to make them appear black.

The government countered these abuses by issuing "propagation permits" to new farmers. These inexpensive permits allowed capture in any season of a specified number of animals for a specific farm. In contrast to farm-reared foxes, no animal captured for propagation could legally be pelted or be shipped live out of Alaska. These two restrictions were directed at known fur farming scams. In one scam, a trapper would masquerade as a farmer, obtain a permit, and trap out of season. He would hold his "breeders" just until winter. Then when the fur was prime, he would pelt all the animals on his "farm."

Another even more lucrative deception was to capture foxes under a propagation permit for a phantom farm and then promptly ship the live animals (with an affidavit claiming that they had been born on a farm) to the state of Washington, where fur farming was just getting started and silver fox breed stock sold for $75 per animal.[38] As one government biologist noted, these "malefactors" were essentially trappers taking animals during a closed season and illegally removing the animals from Alaska's wild stock. True farmers, on the other hand, were "benefactors" who increased the fox population.[39]

The Pioneer Period 1867–1909

"T. V. Smith's catch, Middleton Island, Alaska, 71 blue fox skins, value about $2000.00." Photo by P. S. Hunt taken between 1904 and 1912. Value in current dollars approximately $50,000.
Mary Whalen Photograph Collection, UAF 1975-84-77, Archives, University of Alaska Fairbanks.

Genuine farmers with propagation permits usually located fox dens in the spring. When the pups were old enough to venture out, the farmer would trap them at the entrance by using modified leg-hold traps whose springs had been weakened by heat and whose jaws were thickly padded with cloth. It was exacting work. The farmer had to set the traps in the early morning before the vixen returned from her night hunt, then hover in hiding to quickly transfer trapped pups with desirable silver coats into a cage. A more indiscriminant and destructive approach was to dig out a den before the pups were old enough to run. Although legal, this method was decried by government officials because it destroyed the burrow, stranding the red pups that farmers left behind.[40]

Despite challenges of getting started, a number of Alaskans were keen to raise silver foxes. In the first few years of the twentieth century, news had reached Alaska of a fortune being made by two Canadian farmers on Prince Edward Island. A few years after the Semidi Company began farming blue foxes, these Canadians, Charles Dalton and Robert Oulton, had their first success in developing a purebred line of silver foxes raised in pens. After several years of careful breeding, the dark foxes had multiplied sufficiently to allow Dalton and Oulton to quietly collect handsome returns from selling pelts. Then word leaked out that they were farming rather than trapping. At first, Dalton, Oulton, and three other Prince Edward Island fox farmers stuck to a pact to corner the pelt market by selling no breeders. For nine years, the five farmers resisted the demand for live animals. When the pact broke, the five quickly became wealthy selling foxes to eager new farmers.[41] Breeders sold for as much as $8,000 per pair. One legendary pure black fox pup sold for $9,000.[42] In the years leading to World War I, silver fox farms proliferated in Canada, New England, and across the northern tier states.[43]

Attempts at silver fox farming in Alaska prior to 1900 typically failed. The earliest effort may have been that of Captain Otto Carlson, an Alaska Commercial Company employee, who in 1880 tried without success to get silver foxes to breed in pens at Sand Point.[44] As mentioned, M. L. Washburn's early attempt to raise penned silvers also failed. Several years later, Washburn did succeed in shipping some Alaska silver foxes to Maine for a Semidi Company experiment in East Coast island farming. Despite their reputation as poor travelers, the foxes arrived in good condition and were freed on an island tended by two locals. Within a few years, based on tracks crisscrossing the snow, the foxes appeared to be thriving. Unfortunately, they evaded live traps so successfully that soon the Semidi Company could no longer obtain any "definite information concerning them," and the experiment was abandoned.[45]

Other fur farmers tried to raise free-range silvers on Alaska islands. At least eighteen islands between Prince William Sound and Kodiak were stocked before World War I.[46] However, on most of these islands the silver foxes were removed within a few years by trapping or poison and replaced by more tractable blue foxes. Commentators blamed the animals, describing the silver foxes as "suspicious, nervous" or "ferocious and much given to killing

The Pioneer Period 1867–1909

Silver fox reaching for a tidbit inside large chicken wire pen on Sukoi Island, 1919. Photo by C. H. Flory, courtesy of U.S. Forest Service, Alaska Region.

their young." "The wild, savage nature of this animal made domestication unprofitable."[47]

But silver foxes pen-reared on Alaska's mainland did much better than free-ranging silvers on islands. Because foxes are accomplished diggers and climbers, pens had to be constructed from heavy wire netting. A typical pen for a pair of foxes was 10 × 12 feet with 10-foot wire walls extending 3 feet underground and capped by a 2-foot overhang. The pens opened into a common exercise yard, similarly fenced and about a quarter-acre in size.[48]

A farm at Tolovana Hot Springs had significantly larger pens—thirty that were 6 × 27 feet and another thirty that were 27 × 27 feet. The owner, George L. Morrison, had been working as a trader in Tanana in the early years of the twentieth century when he bought three wild silver fox pups from Chief Titus. It took several generations before Morrison was able to develop a consistent breed of silvers. By that time, he had moved his trading post and fur farm upriver to Tolovana. Government inspectors and travelers such as missionary Hudson Stuck took note of the isolated thriving farm.[49]

In later years, Morrison would become one of the best-known fox breeders in the United States and Canada.

In contrast to coastal Alaska where farmers relied on marine foods, Interior Alaska farmers fed their foxes river salmon, wild rabbits, and large game. A government report lists a year's worth of food for one farm: "1½ tons of dried salmon, ½ ton of dried moose meat, 1 ton of rice and 2,700 rabbits."[50] The food was preserved in permafrost tunnels or root cellars. No leases were available. Most inland farms were established by squatting—erecting buildings and living on a parcel of land to gain ownership that was acknowledged by locals if not by the U.S. government. A 1915 map of fur farms shows that many farms in the Interior were solitary compounds located on a river bank—one on the Kobuk, one on the Kantishna, two on the Porcupine, four on the Tanana, and eleven on the Yukon.[51] A cluster of fur farms had formed around Fairbanks, Alaska's largest town of the period. More surprisingly, several were grouped around the tiny settlement of Copper Center on the Richardson Trail.

Charles Heideman: Fox Farm Promoter at Copper Center

Fur farming in the Copper Center region had been encouraged by an early government employee, Charles W. H. Heideman. Heideman was in charge of the Copper Center Agricultural Experiment Station from 1907 to 1909. Subsequently, he started a silver fox farm on nearby Tazlina River. He was an enthusiastic innovator and a promoter of his own scientific theories and business schemes. Although his formal education had ended with elementary school, he became a self-trained agronomist. In his home state of Minnesota, he worked as a railway express agent, served as a state legislator, and published papers on growing fruit trees. Then in 1906, at age 49, Heideman came to Alaska as a sergeant in the Army Signal Corps. When he arrived at Fort Gibbon on the Yukon River, he sent specimens of nine species of grass to the Smithsonian Institution's collection. After Fort Gibbon, his story becomes murkier.

Later, Heideman boasted in a letter to a member of Congress that in 1906 he had undertaken an expedition with only an Indian guide to travel from Fort Gibbon to the Arctic Ocean and back "under the patronage of President Roosevelt." He claimed to have located a million and a half "lost fur seals" from the Pribilof Islands. The drastic decrease in Pribilof seals had not been caused by the pelagic hunting that the government scientists "kept harping on," according to Heideman. As an "expert silver fox breeder," he stated that he realized that the fur seals had been driven to take up residence on islands in the Arctic Ocean because their breeding and whelping had been disturbed by the government seal hunt. He was reluctant to reveal the latitude and longitude of the island rookeries he had discovered because he was hoping that a news organization might reimburse his expenses for a scoop. Unfortunately, all his confirmatory photos and botanical specimens had been lost on his trip between Fort Gibbon and Copper Center when the SS *Leah* sank at the mouth of the Yukon River. He further claimed he had made a second confirmatory expedition to the Arctic Ocean in 1909 and discovered that the fur seals had increased to two million. Heideman's letter was forwarded to government scientists, but neither they nor other travelers to the Arctic Ocean ever found fur seal rookeries.[52]

Heideman had actually gained his experience in fur farming after he arrived at Copper Center in 1907 to take charge of the local Agricultural Experiment Station. He had quickly become intrigued with the silver fox farms in the vicinity. Between plant experiments and reports to the U.S. Department of Agriculture on varieties of local hay for the horses of the Richardson wagon road, Heideman researched fox farming, raised a few foxes, and wrote *A Monograph of the Silver Fox*. This publication reflected the optimism of the day. Heideman noted that the value of silver pelts was due to their rarity, but he was confident that "it will be long before artificial [fur farm] production will seriously affect the price." He was also sanguine about fashion: "Sentiment may condemn the wearing of dead birds to gratify human vanity, but sentiment will never condemn the wearing of furs for comfort." He saw a happy future for Alaska's stockmen, although not in nursing sheep and cattle through icy winters. "Alaska can, and in time will raise the world's fur supply."[53]

When the Copper Center Experiment Station closed in 1909, Heideman turned to his new enterprise, the Alaska Fur and Silver Fox Company, which consisted of a single nearby farm on the Tazlina River. He authored the company's 1910 promotional booklet, *The Story of the Silver Fox*, which contains a third-person description of the expertise of the company's vice president, "Professor" Heideman, who was supposedly a graduate of the University of Wisconsin with many years of experience in fox farming. The booklet was designed to sell 1,200 shares of stock at $25 per share—a wonderful bargain because in fur farming "there is no limit to the market . . . no limit to the food supply. . . . Nature does all the work . . . the risks are minimum and the profits phenomenal." Moreover, potential investors were assured that the Copper Center region was the ideal place to raise furbearers because it already marketed "more high priced silver fox skins than any region of similar size in the world." It was a place where "enough salmon can be caught by one man in a week to feed a hundred foxes for a year . . . where land can be had for the occupation, no rents to pay, no insurance, no taxes, [and] timber for buildings and fuel costs nothing."[54]

Six years later, however, the Alaska Fur and Silver Fox Company had a different view of the Copper Center region. Shareholders received a letter commending them for their patience and announcing that the company had, at a cost of $12,000, finally developed from wild stock two pairs of silver foxes plus several cross foxes. They anticipated two litters of pure silvers the coming summer and were moving the farm to the state of Washington to be close to buyers, because it was "practically impossible to induce those inclined to engage in fox breeding to go to Alaska." Current stockholders were offered additional shares at $25 per share; new shareholders would be charged $50.[55] By this time, the company's former vice president, Charles Heideman, had apparently already left the company and moved to Idaho. Although the company did not prosper, several other silver fox farms near Copper Center continued to operate over the next three decades.[56]

Heideman's silver fox monograph of 1909 roughly corresponds to the end of the pioneer fur farming days in American Alaska. The Semidi Fox Propagation Company was about to fold. Farmers new to the business were beginning to experiment with mink, marten, muskrat, and non-native furbearers such as raccoon and skunk. They would soon be receiving bulletins from a little-known federal agency, the Bureau of Biological Survey, which was working on the science of raising animals for fur. Nationally, the conservation movement, supportive of fur farmers, was gathering strength. One of its champions, Teddy Roosevelt, was already in the White House, withdrawing thousands of acres of public land for national forests, parks, and wildlife reservations. These withdrawals included the Aleutian Islands and long arcs of coastal land in southeast and southcentral Alaska—prime fur farming country.

CHAPTER THREE

Prewar Expansion and WWI, 1910–1918

Conservationist Support for Alaska Fur Farms

In November 1900, the newly elected vice president of the United States, Theodore Roosevelt, was looking forward to reading the report of the Harriman Alaska Expedition. He shared conservationist leanings and personal friendships with several members of the expedition, including John Muir and C. Hart Merriam, whose views would eventually aid fur farmers in Alaska. But by the time the multivolume account of the Harriman Alaska Expedition was published, Teddy Roosevelt had other matters on his mind. Roosevelt, like M. L. Washburn before him, had suddenly been promoted from second to first in command by the murder of his boss, President William McKinley.

Within a year of becoming president, Roosevelt began using his power of administrative order to withdraw blocks of public land for national wildlife refuges, parks, and forests. In August 1902, he issued a proclamation that created the Alexander Archipelago Forest Reserve in southeast Alaska. Alaskans, including fur farmers, were alarmed. Would this mean they could no longer clear-cut timber near tidewater, build salmon canneries, or raise foxes simply by squatting on the islands of Alaska's Panhandle?[1] Powerful naturalists, notably John Muir, spoke strongly to the president about preserving the untouched wilderness within the withdrawals. But other voices, especially that of Gifford Pinchot, who headed Roosevelt's new U.S. Forest Service, spoke for controlled use: "The first great fact about conservation is that it stands for development . . . the recognition of the right of the present generation to the fullest necessary use of all the resources with which this country is blessed."[2] Timberlands, stated Pinchot, should be managed for "the greatest good of the greatest number for the longest time."[3]

Forest Service Fur Farm Permits

In creating national parks, President Roosevelt heeded Muir, but in founding national forests, he listened to Pinchot. In 1907, the president established both the Chugach National Forest in southcentral Alaska and the Tongass National Forest in southeast Alaska. The latter incorporated the Alexander Archipelago Reserve. After further additions, the two new national forests covered 22.5 million acres, equaling 30,800 square miles, which the Forest Service was instructed to manage for multiple uses.

In Alaska, the Forest Service briskly organized a permit and fee system for fur farms inside the Tongass and Chugach National Forests.[4] Rather than adopting the Treasury Department's policy of charging an annual fee of $100 per island, the Forest Service based fees on acreage and on the availability of desirable features such as reliable fresh water, a sheltered harbor, and

Tongass fur farmers W. H. Dugdell and Hardy Trefzgar, Yakutat area, 1926. *Photo courtesy of Mary Graves Zahn.*

Prewar Expansion and WWI, 1910–1918

Chugach fur farmer with harvest of ninety-eight blue fox skins from Lone Island near Cordova, 1925.
Photo courtesy of U.S. Forest Service, Alaska Region.

areas suitable for fox dens. But the Forest Service's biggest innovation was an introductory fee. For the first five years of a permit, every farmer was charged just $25 per year.[5] This eased the difficult start-up period when new farmers earned nothing while expending capital on buildings and equipment and on nurturing a few breeders into a sustainable herd. In the sixth year, forest rangers measured the acreage of an island and assessed how closely its features approached "100% desirability." For an "80% desirable" island, the farmer's fee of 15¢ per acre was reduced by 20 percent. The minimum yearly fee was $50. The maximum was $375 because under Forest Service rules islands larger than 2,500 acres (4 square miles) were not eligible for permits.[6]

Legal access to southeast Alaska islands, the five-year flat-fee incentive, and strong fur prices in the first decade of the twentieth century encouraged many enterprising Alaskans to start coastal fur farms. Not all of these pioneers chose to raise foxes. Alaska's fur farmers tended to have an experimental bent and an outlook akin to that of gold-fever–infected prospectors: Each new rock outcrop or novel fur might be the next bonanza. Early fur farmers were encouraged by a relatively new agency in the Department of Agriculture—the Bureau of Biological Survey—that was just starting work on the science of raising a variety of furbearers.

The Biological Survey and New Farmed Species

The first chief of the Biological Survey was C. Hart Merriam, friend of Theodore Roosevelt and member of the Harriman Alaska Expedition. Merriam was deeply concerned that loss of habitat caused by human population growth would doom many birds and animals to extinction. Already the American bison was in danger, and the passenger pigeon was past recovery. The Biological Survey, founded in 1886, sent expeditions through Mexico, Canada, and the United States to collect data for the creation of "biogeographic" maps that outlined what are now called *ecological zones*. These expeditions, including four to Alaska, used a significant portion of the agency's budget. Merriam had to be reminded by his bosses in the Department of Agriculture that the agency was expected to emphasize economic development more than scientific research. In fact, the original name of the bureau had been the Division of Economic Ornithology and Mammalogy.[7] Merriam deftly proposed to study a subject on which his agenda and that of his bosses coincided—fur farming. Fur farms would help spare wild furbearers and at the same time strengthen the economy.

The Biological Survey began experiments with mink, marten, foxes, fishers, badgers, and skunks at the National Zoo in Washington, DC, and on farms near Pritchard, Idaho, and Keeseville, New York. In Maryland marshes, survey agents studied the rearing of beaver and muskrats. In 1908, the agency published its first bulletin of advice for fur farmers—*Silver Fox Farming* by Wilfred Osgood. This was followed two years later by a manual on muskrat farming. Then, in the 1920s, came a steady succession of bulletins on raising blue foxes, beaver, rabbits, skunks, and mink.[8] In 1909, a nongovernmental author, A. R. Harding, added to the literature with *Fur Farming*, a book that offered advice on raising all of these animals as well as opossum, raccoons, marten, and otter.[9]

How much of this early written material reached the Alaskans who were trying to raise species other than foxes is hard to gauge. But it is clear that with or without books and bulletins, pioneer fur farmers pushed ahead with ingenuity and optimism. They captured breed stock from the wild. They designed pens, dens, and feed mixtures based on often rudimentary understanding of the animals' habits and habitat. Their efforts drew the attention of news reporters and government agents who recorded both marginal successes and speedy failures.

For example, novice mink farmers near Haines created pens with small streams running through them. The mink liked the water and seemed to adapt easily to the pens. But inducing the mink to breed proved difficult, and yield was too low for profit.[10] Mink was not yet a high-end luxury fur. Because the narrow mink pelts required hundreds of time-consuming seams to create a coat, the fur was used mostly for trim, muffs, or small garments. With the price at $4.50 a pelt, mink farmers needed thriving litters if they were to have any chance of making money.[11]

Several Alaskans tried to raise marten but met with little success. Marten are light-colored relatives of the valuable Russian sable, and their pelts sold at nearly twice the price of mink. Unfortunately, marten breeding habits were a mystery that would not be understood for more than a decade. (As late as the mid-1920s, a farmer near Anchorage was so convinced that his lone female marten had been succesfully impregnated by the family cat that it was reported in a fur farm journal as a scientific advance.[12]) Penned marten near Tolovana, Juneau, Copper Center, and Sitka all failed to reproduce. The only reports of fruitful mating came from two southeast Alaska farmers. Bert Maycock loosed marten to roam free and provided supplemental food on Windfall Island. John Fanning did the same on Etolin Island. When spring came, both saw baby marten, although they never identified a breeding season. Nevertheless, both farms closed. Maycock found marten less profitable than foxes, and Fanning's martens disappeared, possibly trapped by poachers.[13]

A man named Rufus Blakely near Ketchikan and an Aleut whose name is not recorded each attempted to farm an animal that had no difficulty breeding—the muskrat. Unlike mink and marten, the muskrat is a rodent and capable of raising two or three litters of five to ten young each year. Although the price for a muskrat's utilitarian pelt was only 30 cents, these farmers were counting on volume to make a profit. Six breeding pairs could theoretically be the nucleus of a herd numbering 400 to 600 by the end of one year if the habitat suited them and predators were kept at bay. For Blakely, the habitat turned out to be the problem. He could not get aquatic plants to take root and flourish in his lake on Bold Island. The Aleut took on an even more challenging task: transplanting a dozen muskrats from a swampy forest on the mainland to a treeless Aleutian island in 1913. Whether it was this major change in habitat or the onset of World War I that scuttled his experiment is not known, but the trial was apparently short-lived.[14]

Perhaps the most exotic experiment during the pre–World War I period was an island skunk farm off huge Prince of Wales Island that marks the southern border of southeast Alaska. In the early 1900s, skunk pelts sold well in London. The most valuable natural pelts, those with the least white hair, brought about $3. Pelts with the white stripes obliterated by dye and renamed "Alaska sable" or "black marten" sold for considerably more. Skunks were not native to Alaska, but they were common in agricultural counties further south where they were known to be surprisingly tractable and good mousers. As early as the 1880s, a few crop and dairy farmers had begun to raise skunks as a sideline.[15] Most of these farmers removed the scent glands while the animals were young. But when William Forbes developed his herd of fish-fed skunks on an island in Shakan Bay off the outer coast of Prince of Wales Island, he simply avoided being sprayed by treating the animals gently. The *Skagway Daily Alaskan* reported that the ship captain who brought the skunks to Alaska did so only with the express agreement that they could be thrown overboard if they proved disagreeable.

The reporter traveled to the skunk farm island and described the scene: "When Forbes and I landed on the beach at their feeding ground, he began calling and hundreds of the animals ran out of the forest. I stampeded, but Forbes fed them . . . fish. They were as playful as kittens." But hundreds of skunks were not enough for Forbes. He was waiting to export skins and skunk oil (used as a chest liniment to cure colds, an emetic to cure whooping cough, or a lubricant for leather) until he could ship 1,000 pelts per year. He anticipated that he had only three years to wait.[16] Whether he made his fortune is unknown. The newspaper did not print a follow-up story, and the skunk farm is not described in government reports.

Of all the pre–World War I experiments, the hardest to credit is described in a 1900 *Scientific American* article on fur farming in Alaska. According to the author, "On one island bears are being raised, and the proprietor of the bear range has a dozen or more animals."[17] It is possible that the writer was taken in by someone's exaggeration of the lifestyle of one of Alaska's "bear men" who live in proximity to human-tolerant bears. Not only the safety but also the economy of a bear farm is doubtful. Bears multiply slowly and pelt prices for top-grade bear skins in 1900—$10 for black bear and $6 for brown bear—did not compare favorably with $60 for silver fox or $25 for blue fox.[18] Although this report seems farfetched, pioneer fur farmers in Alaska were innovators, and it is likely that a variety of plausible and implausible experiments were attempted.

Specialty farmers and blue fox farmers alike were attracted to the mild climate of southeast Alaska. The Tlingit and Haida Indians of the region began to feel crowded by a permit system that excluded them from their traditional use of local islands. Another decade would pass before the Indians officially protested, but from the beginning, relations between Tongass fur farmers and Tlingits were uneasy and sometimes hostile. A farmer might be angered by Indians who hunted deer on his island, discharging firearms during the delicate period when foxes were having pups. Indians might be equally angry to discover their island smokehouse had been converted to fur farm storage. However, unlike conflicts between Indians and farmers in the American West, there was little outright violence. Fur farms continued to multiply not only in the lush Tongass Forest but also in the Chugach Forest, where blue fox farmers on islands in Prince William Sound and silver fox farmers on the Kenai Peninsula both received Forest Service permits.

Bolshanin, Applegate, and the Aleuts

By contrast, the spread of fur farming in the treeless Aleutians, where the Russians had first raised foxes, was slower. Although local Aleuts were more likely to join the new industry than to resist it, weather and geography made hauling supplies and gear to unpopulated islands challenging and costly. Moreover, government administration in the region was fluid and at times confused.[19] During his presidency, Theodore Roosevelt had named a few Aleutian Islands as wildlife refuges. His successor, President William Taft, on his last day in office in 1913, created the Aleutian Island Reservation. The boundaries enclosed the entire chain of islands from Unimak to Attu. Taft's executive order specified that the reservation was to be "a preserve and breeding ground for native birds, for the propagation of reindeer and fur bearing animals, and for the encouragement and development of the fisheries."[20] Taft and his advisers failed to understand that adding non-native, carnivorous animals to more and more islands in the Aleutians would threaten the existence of ground-nesting sea birds. At the time, these migratory birds, numbering in the millions, seemed limitless.

Like the Russians, most of the Americans in the Aleutians were raising blue foxes. By the end of the nineteenth century, a score of men from various states had established island farms. Most of them brought wives with them or married locally and settled in the region to raise their families. They witnessed a series of changes—the virtual disappearance of the sea otter, the start of $100 fur farm leases issued by Treasury agents, the spurt in breeder sales as blue fox farming spread, and the worsening economic condition of the Aleuts as the sea mammals on which they depended for food and clothing succumbed to commercial hunting.

The near extinction of the sea otter and the 1911 international ban on hunting this furbearer caused special difficulty for the Aleuts. Since Russian times, sea otter pelts had been an important source of Aleut income. After sea otter hunting was banned, many Aleuts had difficulty entering the cash economy. In this setting, two men—Nicholas Bolshanin, an Aleut deputy customs agent for the U.S. government, and Samuel Applegate, an early fox farmer—separately concluded that fur farming was the most practical way for local Aleuts to earn money. Certain barriers had to be overcome—the cost of leases and the availability of breeders, for example—but no one knew the islands, the weather, and potential sources of animal feed better than the Aleuts, some of whom already managed fox islands for absentee owners.

Applegate, originally from New Jersey, had arrived in the Aleutians in 1881 to take charge of weather stations for the U.S. Signal Corps. When it came time for him to rotate back to the States, he resigned his post and remained in the Aleutians to trade, hunt sea otters, and eventually to farm blue foxes.[21] In 1894, U.S. Treasury agents on the Pribilof Islands agreed to sell him about a dozen young foxes that he could raise as breeders. But the blue fox populations of St. Paul and St. George were at a nadir. Two years and many letters later, Applegate finally received thirteen foxes.[22] By 1899, he had settled his increasing herd on Samalga Island. For the next fifteen years he made lease payments from $100 to $200 a year, first to the Treasury Department, then to the Department of Commerce, and ultimately to the Department of Agriculture as bureaucratic responsibility for the islands shifted and lease prices rose.[23]

In addition to his fox farm, Applegate owned a trading post and a schooner from which he mapped stretches of Aleutian coastline.[24] He was a man who clung to his east coast origins. His house at Unalaska had gingerbread trim; and regardless of the type of work he was doing, his daily outfit was a starched white shirt with a high collar and a morning coat.[25] He wrote long letters to government officials on assorted subjects—one of which, composed in 1914, protested an increase in yearly fur farm lease fees from $100 to $200. The letter, addressed to the Department of Commerce, describes how close to the margin most of the fox farmers in his region were. He lists ten islands abandoned by prior leaseholders. He offers as an example his own balance sheet for the previous twelve years. He considers himself more successful than most and possessed of money-saving advantages—his island was rich in natural foods; he hired reliable and inexpensive Aleut caretakers; and he used his own schooner for the farm and other enterprises. In the previous twelve years he had sold an average of forty-eight pelts a year at $47 a pelt—a yearly income of about $2,230. But after farm expenses (including his lease and the 10 percent London auction commission), his yearly profit totaled only $305.

By increasing the lease fee, Applegate contended, the Commerce Department was failing in its duty to encourage new enterprise. "The high rentals charged will prevent many poor men from taking advantage of the privilege offered [to start a fur farm]. Men with considerable money to invest are not going into any such business as this, if they know the facts, as they can invest their money to better advantage elsewhere." His proposal for rolling back lease fees suggests he was aware of Forest Service policies in the Tongass and Chugach National Forests. Applegate's recommendation included a flat rate of $25 per year for the first three years with fees thereafter to be based on island size.[26]

In 1916, Applegate wrote another letter of advice to the government. His subject was Aleut fox farmers. As was the case with many white immigrants, his racial attitudes were complex,[27] but he recognized the Aleuts' long-standing use of the islands and was troubled by the economic woe created by the decline in sea-going mammals and fish. He derided the Department of Agriculture's belief in the myth of "unused" islands. The department had invited new fox farmers to take up "any island not being used by natives." "I would like to know," Applegate wrote, "how this is possible to go on an island that a native cannot claim is used by them." He opined that he would have to wait for a volcanic eruption to produce an island then rush to it before the first Aleut got there.[28] He supported a government decision that Aleuts should be allowed to claim islands for fox farms at no charge and recommended that Aleuts and white men married to Aleuts be given the first choice of islands.

Applegate's advice was offered in response to a question that the Bureau of Biological Survey had posed to several prominent men in the Aleutians. Should blue foxes be taken from Attu Island to supply new Aleut fur farms near Unalaska? Applegate expressed alarm about robbing Peter to pay Paul. Attu Natives, in his opinion, already had too few foxes to support themselves. The government should be helping Attu villagers transfer some of their stock to nearby Agattu Island to expand their own herd. The Biological Survey should supply Pribilof foxes to Aleuts on other islands "as cheaply as possible, while white men should pay more." He wrote in the belief that Pribilof Natives were "well cared for by the government and it would not matter to them how many foxes are taken from their islands."[29]

The Biological Survey's question to Applegate about taking blue foxes from Attu was occasioned by a request from a vigorous promoter of fox farming among Aleuts, Nicholas Bolshanin. Bolshanin was an Aleut born in Sitka and schooled in both Russian and English. He was already 30 years old and a widower in 1905 when he moved to Unalaska as deputy collector of customs. Although he had lost part of his right hand in a fireworks accident on New Year's Eve in Sitka, he wrote letters and customs records in clear script. Unalaska was his fourth assignment with the U.S. government, and he greeted a return to his parents' region with enthusiasm. For the next two decades he remained in Unalaska, where he remarried, established a blue fox farm on Unalga Island, and was eventually appointed U.S. Commissioner as well as customs officer.

Bolshanin and a partner stocked their Unalga Island fox farm in 1913. Within two years he was encouraging other villagers to do the same and acting as an intermediary for them with the government. In 1916, he wrote to thank the Secretary of Agriculture for promptly issuing permits for seventeen islands and for considering an additional six. Since most of these islands were operated by partners, the names of some forty villagers appear in the list of permittees. In his letter, Bolshanin states that his goal is to aid Aleuts "to better their condition. . . . In the fox propagation business they shall be on a fair road of recovery . . . in a position to be self-supporting, happy and contented

Nicholas Bolshanin, front left in detail from photo of U.S. Customs Officers, 1906.
Alaska State Library, Portrait File, ASL-Groups-Customs Officials-1.

people." He requests the use of a small government schooner for planting foxes on the villagers' islands.[30] Bolshanin hastens to add his wholehearted support for non-Aleut fur farmers who plan to make the Aleutians their permanent home, but proposes that the government exclude from the islands absentee farmers who reside in the United States and return only during the harvest.

"There is very little here for the native and what little there is . . . should be kept for him . . . furs and fish are the sole salvation of the Aleut."[31]

Not surprisingly, during this attempt at rapid economic change, some of the Aleut permitees were slow to follow through. In 1919, Bolshanin was asked to investigate thirty-five islands within the Aleutian Island Reservation that were held by Aleuts under a fur farm permit. Some of the smaller islands were unnamed and identified only by latitude and longitude. Bolshanin reported that he was unable to get information on six and that another thirteen had never been stocked, although three owners were still waiting for blue foxes to become available from other Aleut breeders.[32] Despite this erratic start, Aleut fox farming took hold especially in the vicinity of Attu, Atka, and Unalaska. The farms continued to provide income to the region for the next quarter century. As for Bolshanin, one writer describes him as "the most successful lessee in the early years of fur farming in the Aleutians," citing the $49,000 he grossed during a three-year lease on Kalvaga Island in the 1920s.[33]

Nicholas Bolshanin and Samuel Applegate agreed on the benefits of fox farming for themselves and for the people of the region, although they differed on many details. Bolshanin had prevailed over Applegate's objection to getting fox breeders for Aleut farms from Attu. Afterward, Applegate had to admit that the Attu herd still seemed robust, although he considered Bolshanin's payment in trade goods penurious.[34] In the end, both men left the Aleutians. Applegate sold his Alaska businesses and moved his family to San Francisco in 1919—one year before the Biological Survey put into practice his proposal for fur farm lease fees in the Aleutians. (For the first three years the leaseholder was charged $25 per year. Thereafter the fee was based on island size, but was "in no case to amount to enough to be burdensome" to the occupant.[35]) Bolshanin moved his family out of Unalaska and returned to Sitka in 1925.[36] The two men left behind the only region of the newly organized Territory of Alaska where Native fur farmers had a significant stake in the industry.

The Territory of Alaska and the Eagle Bounty

The creation of the Territory of Alaska, which in 1912 superseded the old District of Alaska, was politically important to the fledgling fur farm industry. Under the new territorial organization, the governor was still appointed by the president of the United States, but citizens were allowed to elect a legislature with circumscribed powers. Over the years, the Alaska legislature would prove to be a strong supporter of fur farming. The first of several laws passed at the behest of fur farmers was ratified in 1917 during the legislature's third session. The 1917 law—a 50-cent bounty on eagles—pleased fur farmers, but it shook the Audubon Ornithologists Union and the conservationists who worked for the Bureau of Biological Survey.

Territorial lawmakers had become persuaded that eagles preyed heavily on baby foxes and on salmon attempting to spawn, thus harming both the fishing and the fur industries. To lawmakers, the idea of a bounty on eagles was neither new nor repugnant. Alaskans often regarded predators as dangerous competitors. The Territory of Alaska already had a $10 wolf bounty, and there was a history of at least one private eagle bounty. A decade earlier, Samuel Applegate had tried to reduce the estimated 75 percent mortality of his fox pups by paying 25 cents for every pair of eagle claws that locals delivered to him. This private bounty cost Applegate a total of $275 during the ten years that he offered it. Despite the bounty and his cutting down trees with eagle nests, his pup loss continued to be unusually high.[37] No one studied the effect of Applegate's private bounty on the local eagle population. Detractors and supporters of the law had one point of agreement—the new territorial bounty was bound to have a greater impact on eagles because it would apply throughout Alaska.

The eagle bounty issue created a deep division within the Bureau of Biological Survey. At the time, government bounties on predatory or "noxious" animals that threatened farms and ranches—for example, wolves, coyotes, cougars, bears, prairie dogs, and rabbits—were common, especially in the American West.[38] In the past, bounties in various states had also targeted birds—hawks, owls, robins, English sparrows, blackbirds, kingfishers, and starlings. When the 1913 Migratory Bird Act and a treaty with Canada began protecting all nongame birds other than raptors, the Ornithologists Union had cheered. Now it appeared that fur farms, while reducing the trapping pressure on wild furbearers, might at the same time pose a danger to the soaring national symbol. Were eagles actually a significant threat to fox pups? Which species—foxes or eagles—neither endangered, needed more protection? What would be the political effect if federal agents spoke out against a vote of the new Alaska legislature? In the end, support for Alaska's

Eagles, believed to be a threat to fox pups, were bounty animals. This one was killed at a blue fox farm on Sullivan Island north of Juneau, 1928.
Photo courtesy of Mary Graves Zahn.

fur farmers within the Biological Survey and the Ornithologists Union overcame opposition to the bounty.[39] But conservationists across the nation were uneasy about the now-obvious conflict between promoting domestic foxes and saving wild raptors.

Farm Failures: The Semidi Fox Propagation Company

Conservationists were not the only ones feeling uneasy. One reason Alaska's fur farmers had sought anti-eagle legislation was that a surprising number of farms were failing, and it was unclear where to lay the blame. When the largest nongovernment fur farm in Alaska, the Semidi Fox Propagation Company, dissolved in 1914, many speculated on the cause of its collapse. Some suggested that it was due to World War I, which caused European fur markets to close with a crash. (In January 1915, at the London fur auction, the best silver fox pelts sold at $365. By the end of the year, the price was 24 cents.[40]) However, the Semidi Company had begun to fail well before war was declared. During its corporate life, the company sold over 5,000 pelts and over 200 pairs of live foxes. The company's peak pelt year was 1903, when it sold 714 pelts at the lowest price of the century: $8.70 each. Thereafter, prices rose, reaching $47 by 1915, but the number of pelts sold each year by the Semidi Company fell almost steadily. In the five years between 1910 and the close of business, the company marketed only 98 pelts.[41]

Some Alaskans blamed the failure of the Semidi Company on poachers—particularly on North Semidi and on Whale Island, where adult foxes with prime coats were stolen in winter and young breeders were abducted in the spring.[42] Others wondered whether fur farming, normally an individual or family enterprise, could be operated by four men likely to have different approaches to business. Perhaps these semi-absentee owners, who had other jobs and other enterprises, simply failed to pay close attention to business and relied too heavily on caretakers.[43] One writer mentioned the high price of cornmeal and the cost of transportation.[44] Another noted that the natural food of island foxes during the winter—clams, mussels, and small fish from shallow water—was usually exhausted about four years after the foxes arrived. Without a substantial feeding program, foxes were lost to malnutrition.[45] A United Press news article titled "Easy to Raise, Hard to Catch" explained that the company planned "to disincorporate" because "agents had no difficulty raising the foxes, but hungry eagles carried off the fat pups or the foxes were too difficult to catch when they grew up."[46] A government agent, apparently mystified, simply summarized the problem: "After flourishing for several years, the [company's fox] colony began to dwindle until it finally became unprofitable and the company went out of business."[47] No one discussed the Ponzi nature of the company's early breeder sales, which tapered off as the market became saturated with new farmers.[48] Nor was there any thought that fox diseases related to overcrowding might have played a role. Common sense of the time said that well-fed, free-running animals remained healthy and increased yearly.

M. L. Washburn's approach to managing the Long Island fur farm was probably typical of Semidi partners. He had only a rough estimate of how many animals were on the island. He did not record data on litter size, pup survival, or deaths from injury and disease. He also had other pressing business interests and eventually became a distant partner. In 1901, the Alaska Commercial Company, having lost the Pribilof contract a decade before, sold most of its assets to sister companies—its shipping operations went to Northern Navigation Company and its retail stores to Northern Commercial Company. Washburn managed the Kodiak store for N.C. for six years before being promoted to company vice president. He left Kodiak to work at the main office of the Northern Commercial Company in San Francisco. When he died in April 1911, company stores in Alaska were closed until noon and their red "N.C. Co." flags were dropped to half-mast.[49]

W. J. Erskine and the Kodiak Fur Farm on Long Island

To Kodiak residents it must have seemed as if Washburn's absence created a vacuum that attracted another energetic young entrepreneur. W. J. Erskine was the scion of a family with ties to coastal Alaska. His grandfather had sailed for the Russian-American Company, and his father was the chief officer on an Alaska Commercial Company steamship.[50] In 1900, 18-year-old W. J. Erskine made his first trip to Alaska as a purser on a ship captained by his father. He remained in Nome to cashier for the Alaska Commercial Company during the gold strike, and then he continued with the Northern Commercial in San Francisco, advancing quickly to general manager. He was

W. J. Erskine in Kodiak, 1920s.
Baranov Museum, Carolyn Erskine Collection, P368-1-40.

working in the San Francisco office when the devastating earthquake of 1906 struck. That cataclysmic event seemed to solidify his determination to leave the urban, corporate world of the city and return to Alaska as an independent entrepreneur.[51] Erskine established a small company with three ships carrying goods between Seattle, Kodiak, and intervening ports. In 1908, he replaced Washburn as manager of the Northern Commercial Company store in Kodiak. Three years later, when Northern Commercial put the store up for sale, Erskine bought it. In 1914, when the Semidi Company sold its Long Island fox farm, Erskine and three investors bought that too, renaming it the Kodiak Fur Farm.[52]

The Department of Commerce issued Erskine a fur farm license in 1914. This license required him to submit an annual government form that listed buildings, animals, and pelt sales for the prior year. He was told the report form would arrive on one of the autumn mail boats. When no form had arrived by December, Erskine sent the Secretary of Commerce his own report—a twenty-three-page, typed and bound book with colored subtitles, drawings, photographs, blueprints, and a number of recommendations for government action.[53] The document described changes on the island since the Semidi days. Cattle had been removed and penned silver foxes had been added. In addition, the partners were experimenting with treatments for fox diseases, selective breeding, and new foods. They were feeding foxes salted salmon heads from a nearby cannery, offal from cattle ranches, and Belgian hares raised on the island.

At the time of the report, the Kodiak Fur Farm had two dozen silver foxes and a dozen blue foxes in "corrals." Each of these pens, which accommodated one pair, was 23 × 50 feet with 8-foot wire mesh double walls that extended 3 feet underground to slate bedrock and were topped by a wire mesh roof to block escape by determined climbers. The double walls prevented neighboring males from fighting through the fence. Each corral contained a two-level nest box with windows. In addition to the penned animals, the farm was raising forty blue foxes that ranged free but gathered daily at twenty-four feed houses scattered around the island. Breeding records were kept on the penned animals, and all foxes were actively tamed by keepers who rewarded friendly behavior with treats. Along with size, age, and coat appearance, Kodiak Fur Farm workers evaluated gentleness when they selected breeders.

Erskine's report was optimistic about his farm's ultimate success, but he sounded a note of caution for other new entrepreneurs, particularly those planning to raise silver foxes, whose pelts in 1915 sold for $147. He estimated the start-up cost of a new silver fox farm at $8,000 to $14,000. Boats, fish nets, tools, corrals, feed houses, the keeper's residence and salary would require about $4,000. Ten pairs of breeding silver foxes would cost another $4,000 if wild-caught or $10,000 if purchased. Expenses for the first three to four years before any sales were made would probably reach $1,000 per year. Erskine also cautioned that new farmers should plan on the sale of pelts for income rather than on the sale of breeders because the lucrative market for breed stock was fading.[54]

As part of this cautionary tale, Erskine included a history of the failed Semidi farms, and described twenty-three current fox farms near Kodiak. About half these farms had fewer than thirty animals, most of which were blue foxes. Several farms had been occupied by a succession of owners. Erskine listed what he believed to be the reasons for failure—poor fur prices, rough handling of animals, poaching, and negligence. He identified poaching and negligence as the leading causes for the Semidi failure. There is no mention in the report of the 1912 Katmai eruption that deposited fourteen inches of volcanic ash on parts of Kodiak Island; it apparently caused no lasting harm to local fox farms.[55]

The Secretary of Commerce admired Erskine's handsome report but explained that his department had stopped issuing licenses and no longer required annual reports.[56] His letter failed to mention that lease payments on Long Island were two years in arrears. In spite of complaints from farmers such as Samuel Applegate in the Aleutians, the Commerce Department had not yet changed from the 1913 lease price of $200 per year, but the department was lax about collecting or about identifying squatters.[57] (Agents were probably unaware of the existence of many of the small fox farms mentioned in Erskine's report.) In 1919, the government finally notified Erskine that the island farm he had "bought" was actually on land leased from the Department of Commerce. Erskine's purchase price had gained him foxes and improvements only. Commerce officials, recognizing that Erskine was the victim of a misunderstanding, simply started his lease in 1919.[58]

Bureau of Fisheries: New Managers on the Pribilofs

Perhaps one reason the Commerce Department was less than attentive to private fox farmers was that their agents were busy on the Pribilof Islands. In 1910, the government terminated its last contract with commercial companies in the Pribilofs. Management of the reservation shifted from the Treasury Department to the Bureau of Fisheries, an agency of the Department of Commerce.[59] The incoming agents were assigned to provide for the needs of islanders and to supervise fox farming and fur seal hunts. In addition, they had a new duty—to market all skins harvested on the islands. In 1911, Bureau of Fisheries agents imported a herd of reindeer to supplement human and fox diets. They also began shipping blue fox pelts marked with a U.S. government stamp to auction. For the first few years the furs went to C. M. Lampson & Co. in London, but when World War I threatened, the agents switched to Funsten & Bros. in St. Louis, where shipping was secure and freight charges were 15 percent cheaper than to London. Pribilof furs would continue to be marketed in St. Louis for the next fifty years.[60]

Most of the income from the Pribilofs came from fur sealing, but profit from fox farming was also important to the Aleuts. This was not because there was any significant cash to be earned from feeding foxes, harvesting them, and preparing the pelts. The wages were low and unchanging—$5 per pelt to be divided among all fox workers. But the government's net profit from fox pelts was supposed to be used to purchase food for the islanders. Because profits were erratic, so were provisions. During certain winters, islanders suffered actual deprivation. In 1916, when Samuel Applegate wrote to the Biological Survey that breeding foxes needed by new Aleut farmers should be taken from the Pribilofs, Applegate assumed that the islanders were so "well cared for by the government" that a decrease in foxes would not matter to them.[61] But in that same year, Fisheries Agent Harry C. Fassett wrote to the Commissioner of Fisheries and strongly recommended a change from variable "in kind" payments to fair cash wages for workers. According to Fassett, cash on the islands was so scant that the Aleuts could not order clothing or supplies through the mail. And he declared that quantities of food in the government commissary were often insufficient. Nevertheless, the system remained unchanged. In 1917, the government sold 567 blue fox pelts at $61 each for a total of $34,653. The islanders received $2,835 ($5 a pelt) to be divided among fox workers plus "free" food selected by the government and issued in prepacked grocery bags.[62]

Wartime Losses and Profits

In 1917, prices for blue fox pelts were high in St. Louis because, although European fur imports had been curtailed by World War I, American demand for fur had not decreased. The years leading up to that war had been a time of opulent fashion—a period of sweeping skirts, drooping ostrich plumes, and draping furs. Stylish women had stepped out in the evening wearing long, silk-lined, wrap-around capes of horizontal, encircling furs or in cloth coats half hidden by belted fur stoles with tails cascading toward the hemline.[63] Most of the silver and blue fox furs came from farms. Early in the century, furriers, who clustered around the London market and its rival in Leipzig, had been dismissive of farmed furs. But before long, they came to prefer farmed over wild pelts. Farmed furbearers, carefully tended, had glossier fur and fewer scars.

In response to extravagant pre-war styles and a strong export market, Alaska's nascent fur farm industry had roiled with activity—new farms, new species, experiments, failures, switches from silver foxes to blues, and vice versa. When War World I disrupted European train routes and Atlantic shipping, the ostrich feather trade collapsed, and the fur markets of London and Leipzig almost came to a halt. In the United States, fur auctions shrugged off the loss of European fur imports and solicited pelts from domestic fur farmers and trappers. Europe was an ocean away, and American consumers still wanted luxury furs and they still needed utilitarian furs. In Alaska, farmers began shipping furs to New York or St. Louis while trying to cope with a developing labor shortage.

Many island caretakers and young fur farmers in the territory joined other Alaskans in leaving for the military or for high-paying shipyard work in the States.[64] Twelve percent of Alaska's white population, which was largely male, entered either the army or the navy.[65] For the fur farmers who remained in Alaska, the good news was that demand was high, and American fur auctions were rapidly expanding to replace London. The bad news was that many furriers had been trapped on the opposite side of the Atlantic by war. The passage of furs from farmer to consumer was slowed in the garment workshops. Moreover, times were uncertain. No one knew when the war would end or what the world would look like when it did.

Few people in Alaska ventured to start new fur farms. Some farmers, still in the process of building a herd while paying for a lease, watched the international market collapse, lost heart, and abandoned their animals and their islands. Well-established fur farmers were forced to shift from selling breed stock to selling pelts.[66] Some of these farmers discovered that pelt sales alone could not keep them afloat, especially during a period with high labor costs. Others stayed in business by cutting expenses and taking over work usually done by caretakers. On the Pribilof Islands, the U.S. government suffered no labor shortage or unusual costs and profited handsomely from rising prices caused by a relative shortage of pelts. In 1915, twenty prime Pribilof blue fox pelts sold for $255 each.[67]

Several private farmers also continued to send Alaska pelts to market throughout the war. In 1913, George Morrison of Tolovana had been joined by a partner, J. Edgar Milligan, a Canadian trader from Prince Edward Island. Milligan had traveled to Alaska after hearing about Morrison's thriving silver fox farm. One year later, the new partners expanded their business into Canada by transferring fifteen pairs of Alaska silver foxes to a farm on Prince Edward Island. They also brought three pairs of silver foxes from Alberta to Tolovana in order to restock and minimize inbreeding.[68] Wartime profit from Tolovana pelt sales between 1914 and 1919 put the Canadian portion of their venture on solid footing.

In southeast Alaska, the twenty pairs of blue foxes that James York had started with in 1901 had, by 1914, increased to a herd of about one hundred nearly wild animals. York also made money during wartime.[69] In the Aleutian Islands, Nicholas Bolshanin farmed through the war, as did Fred Liljegren in Prince William Sound, W. J. Erskine at Kodiak, and pen-based farmers in Copper Center. Alaska farmers still in business at the end of the war found themselves in an enviable position. When peace returned, the great post-war fur farm rush began.

CHAPTER FOUR

The Fur Farm Rush, 1919–1924

"A Stampede to Take Up Islands"

In January 1920, a boat nosed up to the dock at Cordova on Prince William Sound and unloaded 140 blue fox pelts worth $17,000.[1] The farmers delivering the pelts were Joe Ibach and his wife Caroline, whom everyone knew by her nickname, "Muz." Just before World War I, the Ibachs had taken over the fur farm on Middleton Island where Thomas Vesey Smith had made good before moving south. The Ibachs worked hard to provide high-quality food for their foxes. An experiment in transplanting Belgian hares to the island as a source of "wild" food failed because the foxes finished off the rabbits in one season, but other feeding tactics were successful and the herd had flourished.[2]

In January of the previous year, when Joe and Muz brought in 100 pelts worth $10,000 (and learned that World War I had ended), the news had made the Valdez paper.[3] But their 1920 harvest caused a sensation that traveled well beyond Prince William Sound. Word of $17,000 in a single season spread through the territory. According to Josephine Sather, soon to begin her own thirty-year career as an island farmer, "Fox ranching . . . became the prevailing topic of the day, and where one could find an island suitable for the purpose [was] the main topic of discussion."[4] Another writer commented that the Ibachs' success "started a stampede to take up islands."[5]

As with most stampedes, the Alaska fur farm rush started because a thrilling story spread quickly through a population eager to believe that a man without prior experience could easily "learn the ropes," work hard, and make big money. Shortages during World War I had driven fur prices up sharply. When war was declared in 1914, an Alaska blue fox pelt sold for an average of $47 at American fur auctions. By 1920, the price was $130. The rise of silver fox prices was similar: A pelt that sold for $147 in 1914 brought $170 in 1920.[6] After the war, American men who knew nothing about the fur market or fur farming headed for the territory, looking for work. These war veterans, former Alaskans and newcomers, believed that the Last Frontier was also the land of best opportunity. The War to End All Wars had guaranteed peace. Personal prosperity was now just a matter of hard work, right location, and ordinary luck.

Initially, however, men hoping to make their fortunes in Alaska were delayed by worldwide bad luck—the Spanish Flu pandemic. In October 1918, influenza broke out in Seattle, Alaska's supply depot. During the first week of the month, seventy-five people in the city died. Alaska Territorial Governor Thomas Riggs, Jr., quickly imposed a maritime quarantine. He stationed U.S. Marshals at ports to prevent sick passengers from debarking. Marshals also manned major trailheads, and nearly everyone, even those who appeared healthy, were prevented from entering the interior of Alaska. Alaska Natives were advised to cancel potlatches and ban visitors from their villages. When the SS *Admiral Farragut* attempted to transfer ailing passengers for medical care on land, she was turned away first at Valdez, then at Cordova, and finally at Seldovia on Kachemak Bay before she steamed back to Seattle with one dead and many sick on board. These measures saved a few towns (such as Fairbanks) from the infection, but they were not sufficient to prevent the rest of the territory from being ravaged. By the summer of 1919, the epidemic was spent, but out of a population of 32,000 whites, 150 had died. Eskimos, Indians, and Aleuts suffered more; at least 2,000 out of population of 23,000 died.[7] Between the war exodus and influenza deaths, Alaska's population decreased from about 64,000 in 1910 to 55,000 in 1920.

By the time the Spanish Flu abated and the ports reopened, men coming into Alaska had missed the 1919 fur harvest and the local excitement in Prince William Sound, but the new arrivals joined residents in talk generated by the 1920 winter fur sales. Many newcomers had originally traveled north with the idea of imitating the prospectors, a few of whom had become famously rich two decades earlier. But some of these men began to change

The Fur Farm Rush, 1919–1924

their minds. The price of gold, regulated by the U.S. government, had not been raised for almost a century, and much of the "easy" gold that could be collected with sweat and simple machines had been mined out. Large mining companies that used industrial methods to extract gold from low-grade deposits were taking over. Most Alaska gold miners had become wage earners rather than fortune seekers.

By contrast, fur farming was an enterprise that one or two men or a family could get into with modest capital and, in just a few years, begin to reap great rewards—at least that was what everyone was saying. One had only to study the fur prices of the last decade to see the opportunity. And furs were not a passing fad of a single decade. They had been in fashion in one form or another for centuries. Moreover, the new craze for "automobiling in the open air" had increased the need for utilitarian furs. Wilderness on the continent was shrinking, and traplines could never meet the burgeoning demand. Newspapers and advertisements from fur breeders were enticing. Any man willing to work could be a successful fur farmer. The widely read naturalist Ernest Thompson Seton was particularly reassuring. He stated that a man who could raise hens could certainly raise foxes and "with this advantage . . . a fox requires no more space or care than a hen but is worth twenty times as

Alaska automobile pioneer Bobby Sheldon on the Richardson Road with fur-clad passengers, skis on front wheels, chains on rear, Spring 1915.
Agricultural Experiment Stations Collection, UAF 1968-0004-02888, Archives, University of Alaska Fairbanks.

much."[8] As for the best place to farm, Alaskans noted that nowhere else in the United States was there more land available for new farms or a climate more conducive to luxuriant animal coats than in Alaska.

Fur Farms throughout Alaska

By 1921, the fur farm rush was stirring throughout Alaska. The Biological Survey estimated that at least 225 fur farms were already operating.[9] That estimate included everything from farms of over a hundred animals that required several employees to backyard operations consisting of a few caged foxes whose offspring supplemented the family income. The majority of the farmers were raising blue foxes on islands.[10] Some of these islands were in the Aleutians but many were in southeast Alaska. Fur farm permits within the Tongass and Chugach Forests increased rapidly. In 1921, the Forest Service oversaw about 110 permits within the two forests. Just three years later, the Tongass Forest alone would have more than 200 permit holders.[11] By 1923, *The Alaska Weekly* would claim that "every available island that is suitable for fox raising in Southeastern Alaska [has been] taken up." But the newspaper saw no bar to expansion elsewhere in Alaska:

> Alaska will have a thousand fur farms in the next few years. . . . It is a credit to the Territory and to those who have embarked in this new industry . . . that over $500,000 have been invested within eighteen months without any advance agents or a single space of newspaper advertising. . . . There need not be any fear as to the fur market; there are more people wearing fur apparel today than twenty years ago.[12]

Joe Ibach, of Middleton Island fame, had acted before the best Tongass Forest permits were snatched up. During the rush, he was among the first to apply for an island farm well south of Prince William Sound. In the early 1920s, at roughly the same time that a *National Geographic* writer was preparing a story describing the romantic, isolated life of "Mr. and Mrs. Crusoe" on Middleton Island, Joe and Muz were finding themselves ready for less isolation and more frequent company. By the time "A Northern Crusoe's Island: Life on a Fox Farm Off the Coast of Alaska, Far from Contact with the World Eleven Months a Year,"[13] was published, the Ibachs had already handed over the Middleton farm to caretakers and had left Prince William Sound. They used their 1919–1920 earnings to take out a permit for Lemesurier Island in southeast Alaska and to purchase its meager improvements. Located in Icy Strait, Lemesurier had five times the acreage of Middleton and was within easy boating distance of the village of Gustavus and not far from the regional hub of Juneau.

Lemesurier Island was a known garden spot. Tlingits from the village of Hoonah had picked berries, hunted deer, and planted potatoes on it until a Forest Service permit gave exclusive use of the island to the fur farmer who preceded the Ibachs.[14] The island had a southern cove that provided anchorage during the summer. By 1926, Joe and Muz had built a house on the cove that was lighter and larger than the Middleton cabin. In the cement of the fireplace they embedded a collection of rare coins that Joe had purchased from a Cordova saloon keeper. They built a guest cabin, planted flowers and vegetables, settled a few blue foxes on the island, and began a varied career that included raising foxes, guiding brown bear hunters, and mining in Glacier Bay. They were known for their generous hospitality. Some of the visitors and guiding clients that they entertained (for example, Rex Beach, John Barrymore, and Joe Lippincott) were famous nationally.[15]

In Alaska's Interior, squatters, undisturbed by the federal Land Office, were also establishing new farms. Most of these bordered towns or lined transportation routes, such as the Valdez-Fairbanks Road, the Alaska Railroad, and various navigable rivers. The most remote of the Interior farms was probably that of Daniel Cadzow, the trader at Rampart House on the Porcupine River north of the Arctic Circle near the Alaska–Canada border. Cadzow and his wife Rachel lived in a large, twin-gabled frame house located 500 river miles from the railroad stop at Nenana and 1,100 river miles from the mouth of the Yukon. Explorers Vilhjalmur Stefansson and Roald Amundsen and others visited Cadzow's trading post on their travels. Like most Interior farmers, Cadzow was raising silver foxes. His were descendents of wild foxes he had dug out of dens in 1913.[16] According to newspapers in Auburn, New York, the hometown he had left in 1884 at age 21, Cadzow was a northern success story. In just twenty years of trading and farming furs in the north, he was worth a half million dollars, and he died a wealthy man at Rampart House in 1929.[17]

On Alaska's northern and western coasts, fur farms appeared near Bethel, Unalakleet, Nome, Teller, Kotzebue, Shishmaref, and Barrow. Importing material to make pens was expensive, but these farmers had some advantages. The arctic weather produced thick winter coats, and feed for animals was relatively cheap. The cold, well-oxygenated Bering Sea supported a host of fish, and salmon runs were strong at the mouths of rivers. Moreover, until the late 1930s a fresh reindeer carcass could be purchased for about $5.[18]

In the early 1920s, nearly all of these farms on the northern coast were raising arctic foxes. A few were raising blue arctic foxes, which were rare in the local wild fox populations. George Goshaw, the trader at Shishmaref, imported farm-bred blue arctic foxes from southeast Alaska for his farm, the Hyperborean Ranch. In 1924, Goshaw paid $20,000 for eighty foxes and sent them by steamship from southeast Alaska to Seattle. There, they were transferred to the SS *Victoria*, which carried them to Nome. At Nome, they were loaded onto a coastal vessel. When they finally arrived at Shishmaref, they were described as "weathering the trip well."[19]

Unlike Goshaw, most fur farmers in the far north raised white arctic foxes, the offspring of local wild parents. Frank Dufresne, Bureau of Biological Survey manager for the Seward Peninsula, reported that by 1925, several Eskimo families had made a successful transition from trapping to farming white foxes, bringing in as much as $2,000 a year to supplement subsistence activities. He commented that "the women do most of the work" with foxes.[20] Some 430 miles northeast of the Seward Peninsula, John Hegness, the non-Native manager of the Northern Whaling & Trading Company at Barrow, was also raising white foxes, feeding them whale blubber, seal, walrus, and reindeer meat. Hegness was well known in the north as the first winner of the All-Alaska Sweepstakes, the premier pre-WWI long-distance dog team race.[21] A strong competitor, he once sent a letter to a fur farm trade magazine protesting its statement that George Goshaw at Shishmaref (latitude 66°) had "the most northern fox ranch in the world." Hegness, who operated a white fox farm at Cape Halkett (latitude 71°), claimed, probably correctly, that his was the world's farthest north fur farm.[22]

White foxes were less valuable than blues, but they were cheaper to acquire in the high arctic and their value was on the rise. During World War I, the price of a white fox pelt rose from about $15 to $35. The new price was not entirely due to war scarcity. White fox was gaining popularity with fur dyers who were perfecting their skills. Not only could dyers tint white fur to create specialty colors such as pale grey or rose, some could transform a white fox pelt into a credible blue fox pelt. By the end of the 1920s, blue fox pelts, which once sold for $60 more than white pelts, would be worth only $25 more.[23]

In the Aleutians, by 1923 the Bureau of Biological Survey was leasing more than fifty-five islands to fur farmers. Several dozen more islands were assigned to Aleuts who held free permits. Still others were being farmed without any formal arrangement with the federal government. Eight out of every ten named islands in the Aleutian Chain would at some point be used as fur farms. Nearly all these islands were stocked with blue foxes. Some islands were carefully run by caretakers or resident owners who nurtured their best breeders and provided regular feedings of fish and cereals for much of the year. Other farmers did not attempt feeding or selective breeding; they simply set leg-hold, snare, or deadfall traps at pelting time. Some farmers, like the Russians before them, hauled barrels of live ground squirrels, mice, and voles to islands, hoping to establish new, ongoing sources of fox food.[24] The foxes welcomed all supplements to their winter graze of native plants, carrion, and beach invertebrates. During the summer they energetically searched for eggs and hunted birds, whose population declined as rapidly as the number of foxes increased.

The Biological Survey was in now charge of both the Aleutians and the Pribilof Islands. Fees for fur farm leases assigned to non-Natives were similar to those charged in the national forests. Aleuts, by contrast, were issued free permits and allowed to buy blue foxes from the Pribilof Islands at $88 per pair, a significant savings over the $200 to $450 charged by private breeders.[25] Several Aleut villages formed cooperatives to stock and harvest islands. Belkofski Natives raised silver foxes on three small islands; all the other cooperatives were raising blue foxes. Unalaska Native Community farmed Carlile Island; Attu Village farmed Agattu; Perryville farmed Chiachi; Atka Village Community farmed Amalia and Amchitka; villagers of Alitak ran the Aiaktalik Native Fox Farm near Kodiak; and Umnak Native Village farmed several small islands as well as large Tanaga Island 400 miles to the west. Before the end of the decade, Umnak's cooperative, based in Nikolski, would earn enough money to purchase a large diesel-powered schooner, the *Umnak Native*, and to build a new orthodox church in Nikolski.[26] The predictions of Nicholas Bolshanin appeared to be coming true, as blue fox pelts brought cash and encouragement to distant dots on the map.

Here and there throughout the territory, experiments with animals other than foxes revived. A. F. Piper started a skunk farm in Seldovia. C. E. Zimmerman added skunks and raccoons to his fox farm on Brothers Island. A farmer at Copper Center and a few others were raising muskrats in swampy acreage, planting "muskrat gardens" with carrots and other vegetables to encourage the foraging animals not to wander off.[27] F. C. Webster started (or at least raised money for) a beaver and mink farm north of Talkeetna.[28]

By far the most popular "new" animals were mink and marten. Propagation permits to capture wild furbearers for breeding had been required since before the war, but statistics on permits were not published until 1925. In that year, thirty-two farmers received permits to capture 4 beaver, 27 red and silver foxes, 40 white foxes, 124 marten, and 157 mink. Marten, high-value animals, whose breeding was still not well understood, were usually an experimental addition to an existing fox farm, but mink were often the sole livestock for a new enterprise.[29]

Demand for mink rose as furriers developed new techniques for lengthening and matching the small pelts so that they could be used for jackets and capes rather than just for trim. In the first few years after World War I, Alaska had only a handful of well-established mink farms able to produce breeders. They were raising western mink, the so-called "Yukon" variety, which were popular with furriers because they were larger than east coast mink and had thinner skins, which made coats lighter in weight. These Alaska farms sold "Yukon" breeders—at approximately $200 per pair—to prospective farmers within the territory and to established farmers in eastern Canada and the States who were looking to improve their herds.[30] Not surprisingly, during the rest of the decade mink was the most common animal requested on applications for propagation permits. Each year between 1926 and 1930, an average of 150 permits were granted to fur farmers and about 850 animals reported captured. About 60 percent of the animals captured were either mink or fox. Beaver, marten, muskrat, weasel (ermine), lynx, and land otter made up the remainder. All propagation permits were issued for "not-to-exceed" numbers. Some farmers caught few or none of the animals they requested. Had they all filled their quotas, about 1,650 animals would have been captured each year.[31]

The Fur Farm Rush, 1919–1924

Taxes, Merchants, and the Seattle Fur Exchange

The 1921 territorial legislature saw this flurry of activity as a fiscal opportunity. Noting that fur farming was Alaska's third largest industry (after mining and fishing), the legislators imposed license fees on fur farmers and fur buyers and levied a tax on pelts.[32] Most fur farmers bitterly resented both the intrusion and the $10 annual license fee. One summarized his complaint in a letter to the *Nenana News*: "Many good fur farms in time would be willing to pay . . . taxes, but not now in their infancy. This present law would sure kill the industry. . . . Let us hope this law will be repealed just as soon as possible."[33] Opposition by the farmers and buyers, the remote location of the farms, and the inefficiency of U.S. commissioners who were supposed to collect the money combined to stymie the law. It was repealed in 1927.[34] The territorial government needed tax money but lawmakers also wanted to encourage fur farmers. Alaska needed this boom to replace the fondly remembered gold strikes of the past.

And it was a boom. As with all stampedes, merchants deftly responded to opportunity. New farmers needed everything from lumber and pelting tools to fishnets and outdoor cooking pots. Merchants increased their stock and their advertising. Offerings from these Juneau businesses were typical:[35]

Alaska Electric Light and Power Company	*Britt's Pharmacy*
ENJOY ISOLATION. No need to be isolated from the rest of the world in this age. Reliable radio receiving sets $39.50 and up. Station KFIU Mondays, Wednesdays, Fridays 8–9 pm. Musical numbers, news, weather, and boat schedules.	BRITT'S WORM MEDICINE FOR FOXES Absolutely reliable and safe—$1 per bottle
Goldstein's Emporium FOX RANCHERS!!!!! Having made an extensive study of fox breeding and scientific feeding, we carry a large stock of all kinds of cereal and fishing gear to catch your feed. ALSO groceries and men's heavy clothing	*Winter and Pond* TAKE PICTURES OF YOUR FOXES. USE THEM FOR ADVERTISING Cameras, Film, Developing

J. B. Burford & Company: SYSTEMATIC FARMING. For proper recording of livestock data, use a Corona Typewriter. The Corona Company is putting out a standard keyboard. JUST THE THING FOR THE FARMER.	*Empire Printing Company* POST YOUR ISLAND "FOX FARM—NO TRESPASSING" Cloth Posters—Single copies 25¢ Reductions on lots of 10 or more

Well-established farmers profited by switching from selling pelts to selling breed stock. In 1921, Nicholas Bolshanin in Unalaska sold 172 pelts. In 1922, he sold 152 pelts and held back twenty foxes to sell as breeders at $300 per pair.[36] In southeast Alaska, James York on Sumdum Island decided to retire after twenty-three years of blue fox farming. In 1923, he pelted no animals and sold his whole herd and holdings to the Juneau stockholders of a new company, York Fur Farms. This news was published in Seattle, and although the article did not specify the terms of sale, the implication was that York would enjoy a rich retirement.[37]

Throughout Alaska, farmers with live foxes for sale solicited business from newcomers to the industry. T. W. Lloyd of Seldovia emphasized that his blue foxes from the Teton Fox Farm were not only selectively bred in pens but were also in short supply: "We Have No 'Islands' . . . only pens . . . 12 pairs left for fall delivery."[38] Jenson and Madsen on Cook Inlet between the Kenai Peninsula and the mainland offered to take orders for "Silver Black Foxes, Pups of the Famous Kussillof Strain." These pups, not yet born, were to be delivered to buyers twenty-four months hence. Although Jenson and Madsen did not name their price, other ads suggest that silver foxes were about twice as valuable as blues, selling for $600 to $800 a pair.[39]

Seattle was a commercial hub for the territory and many of the ads from Alaska breeders appeared in a Seattle publication, *The Fur Farmer Magazine*. Established in February 1923, it catered to fur farmers in the Pacific Northwest and Alaska. By 1925, monthly issues were running at least a dozen advertisements from Alaska fox breeders as well as an ad from the Matanuska Minkery of Wasilla—one of the few Alaska mink farms far enough ahead of the game to have breed stock for sale.

Advertisements in *The Fur Farmer Magazine*, March 1925, page 21.

Seattle was becoming an important market for Alaska pelts. Some Alaskans resumed sending furs to London after 1918, but that city could no longer claim to have the greatest concentration of furriers in the world. After the disruption of WWI, many European furriers immigrated to New York. In response, the New York pelt market expanded, becoming second in the nation only to the venerable St. Louis auction, which had been established in the nineteenth century to serve the trappers of the Louisiana Purchase. Although the U.S. government shipped Pribilof furs to St. Louis and some Alaska farmers shipped to New York, transportation issues—speed and cost—made a regional market attractive.

A Seattle fur auction had been established in 1898 because local merchants needed an outlet for furs that they accepted as payment from Alaska customers. In spite of also receiving pelts from western Canada, eastern Siberia, and the Pacific Northwest, this early auction had a shaky beginning and nearly failed after World War I. Then, in 1923, under the sponsorship of the Seattle Chamber of Commerce, it was reorganized as the Seattle Fur Exchange, an enterprise that would prosper.

Among the organizers was an Alaskan, Charles Garfield, secretary of the Alaska division of the Seattle Chamber of Commerce. Garfield had been one of many unsuccessful Klondike gold seekers, but instead of returning home with empty pockets, he moved on to Nome in 1904 to take simultaneously the jobs of collector of customs, manager of the telephone company, and editor of the newspaper. He first became acquainted with the fur trade in Nome. After several years, Garfield moved to Juneau, where he developed a lasting interest in fur farming. (Later, in the 1930s, he would found and for two decades edit a regional fur farm journal.[40]) In 1922, Garfield moved to Seattle to take the job with the Chamber of Commerce. His first task was to help organize the Seattle Fur Exchange.

In setting up the Exchange, Garfield and other Chamber of Commerce employees recruited big businesses, such as the Northern Commercial Company, as investors. The first board of directors included a bank executive and several businessmen who headed wholesale hardware, dry goods, and grocery stores. The logo of the Seattle Fur Exchange was a silver fox perched on a map of Alaska. Within five years, the new auction house was handling more than 600,000 pelts a year, with an estimated value of four million dollars. Many of these furs came from traplines, but from the beginning nearly all of the blue and silver fox, and some of the mink, came from farms.

The first half of the 1920s was a heady period for fur farmers. The fur farm rush created a strong market for high-priced breeders, and fashions created a rising demand for pelts. Movie theaters spreading through the nation featured silent film stars such as Clara Bow, Mary Pickford, Gloria Swanson, and Louise Brooks, all of whom appeared on screen and in fashion photos nestled in long-haired furs. Newsreels captured both Amelia Earhart in her leather aviator jacket with its broad fur collar and the royal family of England waving from balconies in elegant cloth coats trimmed out with fur.

Seattle Fur Exchange logo, ca. 1923–1939.
Image courtesy of the Dederer Family Collection.

The Fur Farms of Alaska

Headlines in Alaska newspapers proclaimed the economic good news: "Fox Ranchers Getting Rich" (1920); "Wealth in Alaska Foxes" (1922); "Huge Profits in Fox Farms" (1923). Honest farmers, merchants, and auction houses were not the only ones who wanted part of the action.[41] Breeders whose "farms" were dug-out fox burrows, self-annointed veterinarians selling nostrums for animal ailments, furriers marketing dyed rabbit coats as "Baltic seal" or "French sable," and scammers promoting stock in prosperous, nonexistent farms all looked to share the bounty. Opportunity abounded for gullible investors to purchase high-priced foxes from a breeder, who then charged a boarding fee to care for them. Each year the investor was to receive a guaranteed-to-be-huge profit when the pups born to his foxes were sold as breeders. *The Fur Farmer Magazine* and others published articles warning potential victims about these blandishments, but government investigators

Buyers examining furs before an auction at the Seattle Fur Exchange, 1920s.
Photo courtesy of the Dederer Family Collection.

and the courts usually did not get involved.⁴² The prevailing opinion seemed to be that a sucker bore the responsibility for his naiveté.

Poaching and the Fox Branding Law

In contrast to mere head-shaking about swindles, response to outright poaching was strong. Complaints about poachers in Alaska had begun in the early days of the industry. In 1908, Thomas Vesey Smith had foxes stolen for which he blamed Japanese fishermen. This crime was sometimes described as "piracy" because it was blue fox farmers on islands who were most vulnerable. An attempt to steal fox or mink from a fenced farm near a town usually created an animal ruckus that brought the resident farmer running. But poachers in a small boat could quietly land on an island under cover of night or on a day when the caretaker was in town picking up supplies. The poachers would set traps or lay out poisoned meat, then collect the animals a few days later. The few land-based marshals in regional towns had little chance of catching the miscreants. The U.S. Attorney General at one point called on the Coast Guard to help "run down poachers," but there was not much likelihood that the revenue cutter would be in the right place at the right time.⁴³

Losses were significant from the very beginning. Second-generation Aleutian fur farmer Ed Opheim reported that in 1907 his father abandoned pen-raising silver foxes on Unga Island and pelted all his animals after suspected poachers repeatedly landed on the island at night. George Schove in southeast Alaska abandoned Brothers Island in 1910 after having his estimated thousand foxes reduced to twenty by poachers. A farmer on Sukoi Island alleged that in the fall of 1908, poachers captured thirty of his blue foxes and took them to a small island where they were penned until their fur was prime. He claimed that the pelts were sold in Wrangell for more than $1,000. He wrote, "The U.S. Deputy Marshal and government officials at Wrangle [sic] . . . gave me no satisfaction whatever." He suggested, fruitlessly, that "any man who leases an island from . . . the federal government should be a good enough citizen to be sworn in as a deputy marshal . . . with the authority to arrest a poacher without having to leave his farm."⁴⁴ Before WWI, several farmers resorted to armed watchmen and private rewards, but most took their chances and hoped that no thieves would land.

After the war, poaching boomed with the industry. Indians, who had hunted and trapped on the islands before the farmers gained permits, were often blamed for depredations. Fishermen, who had easy access to the islands, were also suspected. The territorial legislature passed a law that officially made it a crime to steal foxes, reindeer, or various other farm animals. The penalty for conviction ranged from a $100 fine and a month in jail to ten years in the penitentiary.[45] Nonetheless, poaching continued to be a problem: "It is apparent . . . that organized robbery of the fox islands is being carried out at least fairly successfully. The authorities are doing all in their power . . . but are unable to protect all the islands as it would require a good sized fleet to do so."[46]

In 1923, the legislature passed a more successful anti-poaching law that was designed to prevent thieves from selling ill-gotten blue fox pelts. Commonly called the "Fox Branding Law," it was promoted by the Bureau of Biological Survey and modeled on a Canadian law from Prince Edward Island. In Alaska, it applied only to blue fox farms, whether on islands or inland. Each farmer was required to submit an affidavit describing the farm's location, whether leased or owned, years operated, and the names of all partners. Title and lease information was checked by the Territorial Department of Audit. Then a $10 certificate showing the registered brand was issued to the owner. If a farm changed hands, owners submitted a bill of sale and paid $2 to have the brand re-registered.

The term *brand* is potentially misleading. A three-quarter-inch, triangular arrangement of two letters and one number was marked inside a fox's ear, not with a hot iron but with tattoo ink. The Territory of Alaska obtained from a Chicago firm and sold to farmers a clamp set with tattoo needles forming the letters and number of the brand. The $7 price of a tattoo clamp was never increased during the nineteen years that the law was in effect. The Fox Branding Law also required blue fox farmers to post their farms with signs "within sight of each other," stating "Fox Farm—No Trespassing" in black letters at least six inches high on a white background. The Territory contracted for cloth signs meeting this requirement, which they then resold to farmers. The penalty for trespassing, especially if accompanied by setting a trap or firing a gun, was a fine of $1,000 and/or six months in jail.

Blue fox farmers welcomed the new law. A few had already been using privately developed tattoos—such as the owner's initials or a diamond or a "W" pierced by an arrow.[47] After the new instruments became available in 1924, buyers who purchased or accepted blue fox pelts for auction were required to make sure the brand belonged to the seller. For several years, the territorial government even stationed a deputy game warden in Seattle during the winter auctions to inspect furs.[48] In the arctic, where trappers occasionally caught a wild blue fox, the wild pelt could be sold only after it had been tagged and sealed by the nearest U.S. Commissioner. Between 1923 and the end of the program in 1943, 380 blue fox brands were assigned—85 percent of them during the first two years of the program. By 1925, most blue foxes were tattooed. Judging from the dearth of poaching reports in the newspapers after that date, the Fox Branding Law seems to have worked reasonably well.

The last big poaching story to appear in territorial newspapers, however, was grim. On March 5, 1923, the wife of Ole Haynes, caretaker of the San Juan Fox Farm on Sister's Island, spotted a small boat landing on the far side of the island. It carried two men with a tent, traps, and rifles. Haynes was afraid to confront the men alone, so he took his boat to a nearby island farm and recruited three men to help him. On March 7, the four, concealed by a downed spruce tree, waited on the hill behind the strangers' empty tent. After several hours, a man walked up the beach toward the tent with a dead fox in one hand and a rifle in the other. According to the witnesses, when Haynes ordered him to drop the gun and throw up his hands, the man dropped the fox but held onto his gun saying, "All right boys," before disappearing into his tent. Haynes stated that as he looked down on the tent, he saw the rifle barrel swing in his direction. Haynes fired through the tent wall, hitting Billy Gray in the chest. Gray, an Indian described variously as being from Ketchikan or Metlakatla, died within minutes. One of the witnesses gave Haynes a ride to Pybus Bay, where he caught a cannery boat to Petersburg and reported the shooting. The Coast Guard arrived at the fox farm two days later. Billy Gray's companion and their boat were gone, although two sprung traps were found in the woods. The coroner's jury ruled self-defense.[49]

Alaska's Fur Farm Associations

In facing the coroner's jury, Ole Haynes had the support of the Southeastern Alaska Blue Fox Farmers Association, which had been established two years previously at the encouragement of the Biological Survey. In 1922,

F. G. Ashbrook on left with Clyde Coombs, fur farmer on Perl Island near the Kenai Peninsula, 1923.
Smithsonian Institution Archives, Record Unit 7143, Box 16, Folder 10.

Frank G. Ashbrook, fur specialist with the Survey, had made a four-month coastal tour in the MV *Sea Otter* from Kodiak to Wrangell. Ashbrook had been hired by the government soon after he graduated from Pennsylvania State College in 1914, and he was to have a significant influence on Alaska fur farming during his long career.[50] The 1922 tour was his first trip to Alaska. He was gathering information for what was to become a widely read bulletin, "Blue-Fox Farming in Alaska." In addition, he was bringing scientific and organizational information to Alaska's fur farmers in response to the bureau's mandate to use science to advance the commerce of the nation.

Ashbrook spoke to groups of fur farmers about breeding, feeding, and the advantage of working cooperatively to purchase supplies, to market furs, and to influence legislation. In Cordova, he helped organize the Prince William Sound Fox Farmers Association. In Petersburg, Ketchikan, and Juneau, he helped found the Southeastern Alaska Fox and Fur Farmers Association, later known as the Southeastern Alaska Blue Fox Farmers Association. This association, headquartered in Juneau, chose for its secretary to head the organization Ernest P. Walker, the senior employee of the Bureau of Biological Survey in Alaska. In his letter of acceptance, Walker emphasized his personal as well as official support for fur farming: "My regular official duties give me all, and more work, than I can well attend to but my interest in the success of your organization prompts my acceptance." When Ole Haynes testified before the coroner's jury in 1924, the association had fifty-five members, each paying $15 per year in dues. Farmers in the Kodiak region used the Bureau's model to set up the Southwestern Alaska Blue Fox and Fur Farmers' Association. In Seldovia, the Cooks Inlet Blue and Black Fox Farmers Association also began to meet, issue newsletters, and supply members with reward posters.

The associations offered information on everything from the problem of matted fur to paying income tax. *The Fur Farmer Magazine* began reprinting many of their newsletters. To the magazine's masthead was added the declaration that it was the "Official Journal [of the] Northwest Fox Breeders Association, Inc., the Blue Fox Farmers Association of South Central Alaska, the Cooks Inlet Blue and Black Fox Farmers Association of Alaska, and the Southwestern Alaska Blue Fox and Fur Farmers Assn."

Women Fur Farmers

Membership lists of fur farm associations and territorial records from the 1920s include not only men but also a score of independent women who farmed foxes, mink, and other furbearers. A few, such as Rika Wallen, who raised silver foxes at her roadhouse near Delta, were well known in the territory. Others, including some who were unlisted but equal partners with their husbands, were known only among fur farmers. In 1924, *The Fur Farmer Magazine* published a story about the married daughter of Fred Liljegren, Amanda Bennett, who was described as "one of the ablest blue fox farmers of Prince William Sound." In 1925, Harriet "Mickey" Williamson, "a real woman farmer," was featured for the skill with which she and her husband Billy had raised silver foxes on the Kenai Peninsula for more than a decade. That same year a newspaper also credited Mickey Williamson with persuading the American National Fox Breeders Association to send inspectors to selected Alaska silver fox farms. If the inspectors found the farm clean, the animals healthy, the breeding techniques up to date, and the record-keeping meticulous, the foxes were included on the association's national registry. Having registered foxes was an asset, especially for Alaskans who were selling expensive live breeders to customers who knew them only through the mail.[51]

The role of women in Alaska fur farming is not surprising. Census figures show that an increasing proportion of Alaska's population was female. In 1910, women represented only 28 percent of the total; by 1930, they would reach 40 percent. Social changes also encouraged women to try "masculine" occupations such as fur farming. Women had gained confidence by working while their men went into the military during World War I. The bobbed-hair flapper, gaily flouting convention, was a popular image in the media. And independent women, like independent men, were attracted to and welcomed in Alaska. The initial act of the first territorial legislature in 1913 had been to grant women the right to vote in Alaska elections. Women in many other parts of the nation did not gain this right until the Nineteenth Amendment to the U.S. Constitution passed in 1920.

Several Alaska women published articles about their fur farming experiences. Among these were Josephine Sather and Kay Barker—two women with a similar attraction to the business but quite dissimilar backgrounds.

Mickey Williamson holding prize fox at Williamson Silver Black Fox Ranch on the Kenai Peninsula, ca. 1929.
Photo courtesy of Mary Graves Zahn.

Feeding time at the Hadlund farm on Monte Carlo Island, fifty-five boat miles from Petersburg. Young Katherine wrote to *The Fur Farmer Magazine* (December 1930), "I have lots of pets among my Father's Blue Foxes . . . and a very nice teacher in the person of my mother."
Alaska State Library, Talmage Family Photo Collection, P345-251.

Josephine Sather at her farm on Nuka Island off the east coast of the Kenai Peninsula.
Anchorage Museum, B1991.46.106.

Josephine Sather was a tough Austrian immigrant who, with her second husband, established a fox farm in 1920 at Nuka Island off the east coast of the Kenai Peninsula. She arrived "laden with crates of chickens and washtubs full of plants, . . . a glass-front china closet, [and] a roll of wire netting." Nuka Island was her home for the next forty years. Photographs show her muscular and square-bodied, wearing a gingham dress covered by a full bib apron, a kerchief wrapped around her head, and a broad smile.[52] Kay Barker—described by a Juneau newspaper as a "New York socialite," and a "titian-haired heiress . . . wearing a mink coat"—was half owner of a blue fox farm on Ushagat Island near the entrance to Cook Inlet. In 1930, she and her partner stocked the island with sixty foxes. Barker returned for several winter months in 1937 as part of a three-person crew to harvest pelts and to select twenty breeders as stock for a second fur farm in New York State. She arrived at her island with a movie camera and a boatload of materials for a new cabin with indoor plumbing and an electric-light plant. Although she planned to "return every year" or perhaps even "stay and invest [my] money as [I] see fit,"[53] Alaska newspapers do not mention Barker again.

Different as these women were, their words echo each other. About her island, Barker said, "I reveled in its wild beauty," and Sather wrote, "I was so awestruck with the splendor lying on all sides of me that I simply stood there dumb." About life far from town, Barker commented, "I get so bored with . . . the 'soft' life, I just have to do something for myself"; and Sather celebrated, "I was free of the monotonous ties of routine and responsibility, and my heart was full of enthusiasm." Barker was protective of her foxes: "[When] the fox houses started leaking, I would bring my pets down to the cabin and keep them inside . . . until the storm was over." Sather was maternal: "I put the [orphaned fox] puppy on a pillow by the stove and gave it warm milk at short intervals."

Other women writers describe similar experiences.[54] Taming breeders and nurturing the offspring was challenging but heartwarming. Pelting required setting aside sentimentality, but they took pleasure in their skill and competence. And all of the women describe a strong attachment to the natural world that surrounded their farms. This life was glamorized in the book and movie *Rocking Moon*, in which its beautiful protagonist, Sasha Larianoff (played by Lilyan Tashman), raised blue foxes on Rocking Moon Island in Alaska. Barrett Willoughby, author of the book, based her descriptions on the Kodiak Fur Farm owned by her friend W. J. Erskine.[55] The movie, however, was filmed at a more accessible fox farm on Taigud Island near Sitka. In the story, Sasha is courted by an upright hero and a villainous poacher/mortgage holder—who gives an early clue to his character when he reveals that he prefers the artificial world of southern cities over the natural beauty of Alaska. Sasha is independent but securely feminine, able to keep her white fur coat clean during grubby adventures. She is nurturing toward her foxes, and even keeps a valueless "Samson fox" (a genetic variant with only an undercoat) as a pet. Sasha intends to sell her crop of foxes as live breeders, but when the villain poaches and pelts most of them, she is unsentimentally happy about recovering the pelts and putting them up for sale.

Domesticating Foxes

Regardless of whether feminine influence was a factor, many fox farmers of the 1920s were intrigued with the task of taming these wild animals. The Biological Survey encouraged working toward domestication for practical reasons—the more trusting and calmer the animal, the less likely it was to get into a pelt-scarring fight or to kill its young in panic. As early as 1908, a Bureau bulletin advised farmers to select breeders "with the least aversion to man [which] in time should produce a thoroughly domesticated race of foxes, a result of inestimable value. . . . The success of the business will be assured."[56]

How-to articles written for *The Pathfinder of Alaska* by experienced fur farmers also promoted taming foxes by training. C. E. Zimmerman, who raised blue foxes on Brothers Island 80 miles south of Juneau, emphasized that tame foxes were more prolific. Not only did they not harm their pups, they produced larger litters. One of his favorites, Trailer, at 1 year of age, raised eleven pups, and as a 2-year-old, raised nine. Photos accompanying Zimmerman's article show several blue foxes gathered on his porch, two riding in the prow of his rowboat, one standing on his shoulders, and another relaxing in his wife's arms. His foxes were said to respond to their names and do tricks. Trailer would sit on her haunches to eat out of his hand. Scratcher's trick was quickly learned by the other foxes—paw at the kitchen door and someone will come out and give you a snack. Rags would pull on the Zimmermans' clothing as a signal that he wanted to be petted. In the conclusion to his

C. E. Zimmerman with pet blue foxes on Brothers Island south of Juneau, ca. 1928.
Photo courtesy of Mary Graves Zahn.

article, Zimmerman noted, "Tame foxes are much more interesting to work with than wild ones." [57]

Forest Berry, who reared silver foxes near Homer, was also deeply involved with his animals. Photos in an article by Berry in *The Pathfinder of Alaska* show a fox named Dick Turpin licking Berry's ear and another named Black Oster "kissing" his lips.[58] Berry's enthusiasm for his tame foxes was so great that the year after writing his article, he left Alaska with sixteen crates of foxes to start a fox circus in Wisconsin. He boasted to a reporter, "I have mastered the fox vocabulary and can speak it intelligibly."[59] The fate of the circus is unknown, but Berry's passion for raising foxes apparently persisted. In 1927, another reporter interviewed Berry in California. He and his wife had just driven from Belvedere, Maine, to Hayfork, California, in "an auto-

Tame fox at Williamson Silver Black Fox Ranch monitors a photographer.
Photo courtesy of Mary Graves Zahn.

mobile loaded with $10,000 worth of silver and gray foxes," to start a new fox ranch.[60]

The articles from Berry and Zimmerman about trained and partially tamed foxes raise the question: How difficult would it be to fully domesticate foxes, to develop a breed that from birth behaved like dogs? In the 1950s, a Siberian scientist, Dimitry Belyaev, set out to determine how many generations of carefully selected breeders would be required. He started with 130 silver foxes from an Estonian fur farm, animals already partly habituated to humans. Each year, about 20 percent of the animals were chosen as breeders and the rest were pelted. Coat appearance for both groups was ignored. To avoid confusing behavioral taming with true domestication, none of the foxes received any training. In the first few generations, Belyaev selected foxes that were approachable; in the next generations, those that could be petted; and in later generations, those that were actively friendly, rushing to greet workers with wagging tails. After about thirty-five generations (thirty-five years) of selection, the research station had 100 foxes that were docile, eager to please, and competed for a handler's attention. Some of these descendants also showed physical traits common to certain dog breeds—floppy ears, curled tails, and splotches of white on their chest and face.[61]

Of course, no fox farmer would get so carried away with domestication as to allow white-splotched animals to develop in his herd, but the Siberian experiment does support the farmers' surmise that with many generations of careful breeding, wild foxes would become as easy to handle as dogs. Although not as tame as the experimental Russian foxes, Alaska foxes that adapted to their owners—especially the pups that entertained children and the breeders that stayed on a farm for eight to ten years—were not only easy to handle but also provided isolated farmers with companionship and probably inspired humane farming practices.

Saturated Fur Markets, Animal Disease, and the Price of Feed

In the early 1920s, the most successful fox farm in Alaska was not a one-family island with partially tamed foxes. It was the U.S. government farm on the Pribilof Islands. This operation continued to benefit from inexpensive but experienced Aleut labor and the methods originally developed by James Judge. An official investigator wrote with enthusiasm: St. George is "the most successful fox-farm in the country at the present time, if not the most promising in the world."[62] In 1921, the year after the Ibachs' $17,000 harvest, the government shipped 1,125 pelts to the St. Louis market and netted $108,000.[63] Some private farmers were disgruntled. W. J. Erskine of the Kodiak Fur Farm, for example, complained to the Biological Survey that it was unfair for the government to sell Pribilof fox skins in competition with private fox farmers. The Southeastern Alaska Blue Fox Farmers Association sent a similar protest in the form of a resolution to the Secretary of Commerce.[64] They stated that the government had too many advantages over private farmers: cheap labor, no taxes, no lease payments, and free original breed stock. Moreover, no fox poachers dared approach the distant federal property of the Pribilof Islands.

But by the middle of the 1920s, Alaska fur farmers were worried about more than just government competition. The country was in the middle of an economic boom with a rapidly rising stock market, but fox pelt prices had begun to slip. Blue fox sales were back down to early WWI levels—

The Fur Farm Rush, 1919–1924

Blue fox pelts awaiting shipment outside warehouse at St. Paul, Pribilof Islands. *RG22-95-ADMC-561, NARA at Anchorage, ARC ID 5682401.*

roughly $50 per pelt. Silver fox prices had fallen even farther—to as low as $90 to $100 per pelt. Although some farmers accused others of recklessly dumping inferior pelts from culled animals on the market and damaging the reputation of fox as a high-quality fur, there is little evidence that demand had fallen. Instead, supply had risen. New farms established in Alaska and elsewhere during the post-war stampede now had substantial herds. With demand for breeders down, farmers sold fewer live foxes and sent more pelts to the market each year. Between 1920 and 1925, the total number of blue pelts shipped out of Alaska each year rose from approximately 1,570 to 5,400. The number of silver fox pelts exported from the territory rose more slowly from about 330 to 580.[65] Although Alaska's pelts accounted for only a small minority of U.S. silver fox sales, they made up a significant portion of the country's blue fox sales.[66] The rising volume from Alaska and elsewhere contributed to dropping prices.

Fur farmers in the territory not only had to face falling national prices, they were also worried about problems at home. Island farmers had discovered that crowds of free-running foxes were as prone to disease as penned animals. Those who farmed island foxes and the mysteriously breeding marten complained that the territory had no veterinarians to aid them. Mainland fur farmers, without access to leases, were concerned about land. Nearly all of Alaska was owned by the federal government, and existing law made it impossible to homestead or buy government land for a fur farm. Fur farm leases and permits were available only within a national forest or reserve. With no official standing, Interior and northern coastal fur farmers were at a disadvantage if they wanted to sell a farm, outlaw trespassers, or prevent a homesteader from encroaching on their farm. And everywhere, there was concern about feed. On islands, birds and eggs were decreasing. In the Interior, the number of snowshoe hares rose and fell in cycles, and along the coast, the salmon runs fluctuated unpredictably. Dry fox pellets sold by animal feed companies were expensive, and even shipping cereal into the territory for homemade feed was not cheap. The reality of fur farming was turning out to be more challenging than Ernest Thompson Seton and early promoters had described.

Nonetheless, the industry was robust enough to inspire optimism, and Alaska farmers had confidence in their ability to solve problems. Poaching had already been minimized, lease fees had become reasonable, and regional fur associations were getting stronger. In the next five years, the fur farmers would push hard for veterinary care, cheap animal feed, and land rights. The U. S. Congress, the territorial legislature, Indian advocates, and a new Alaska Game Commission would respond with both opposition and support.

CHAPTER FIVE

The Peak Years 1925–1929

A Territorial Veterinarian and 700 Fur Farms

Of all the issues facing Alaska fur farmers in 1925, legal access to land must have seemed the most straightforward. Nearly all of Alaska's land was owned by the U.S. government, which was already leasing property for fur farms within the national forests and the Aleutian reserve. Fur farmers in the rest of the territory also wanted access to government land. They needed an Act of Congress, but their request was a matter of simple fairness: Shouldn't northern squatters be permitted to secure their land by leasing like the fur farmers of the Tongass Forest, or by homesteading like the crop farmers, or by staking a claim and then patenting land like the miners?

A proposal to change the venerable Homestead Act or alter entrenched mining laws would probably have encountered stiff opposition. Fortunately, the Department of the Interior was willing to draft a bill that would allow fur farmers to lease land administered by the U.S. General Land Office. The bill, "Leasing Public Lands in Alaska for Fur Farming," proposed granting ten-year leases for mainland tracts up to 640 acres in size and for islands up to thirty square miles (19,200 acres) in size. Introduced by Nicholas Sinnott of Oregon, the bill was sent to the House Committee on Public Lands in early 1926. To the surprise of supporters, it met with strong opposition from a previously dismissed group that had been pushed off coastal islands by fur farms—Alaska Natives.

The Alaska Fur Farm Leasing Act

Native groups, encouraged by leaders such as William Paul, a Tlingit lawyer and territorial legislator, were becoming politically active. In 1921, 180 Haida Indians signed a letter of protest to the Forest Service stating that fur farmers had denied them access to small, heavily used islands southwest of Prince of Wales Island. An investigation by the Ketchikan District Forester led to the cancellation of the permits in question. These cancellations were a small victory, however. No other Tongass permit holders were displaced, and the Juneau District Forester's response to conflicts was to recommend that "fox farms keep armed guards on duty, as did fishtrap owners."[1]

When the Alaska fur farm leasing bill was introduced in Congress, the Natives responded promptly. The Alaska Native Brotherhood in Ketchikan wrote "an appeal to Congress for justice to the native people of southern Alaska" that was quoted by New York Representative Fiorello LaGuardia in his opposition to the bill: "From time immemorial our people have subsisted by hunting, fishing and trapping. Many of the islands upon which our fathers hunted and trapped have now been preempted by white men for raising foxes. . . . We have been ordered out of the bays where our forefathers fished." LaGuardia commended the letter and argued in a minority committee report that "it is now attempted to take away from these natives the remaining islands for fur farming . . . and to turn them over to private corporations to exploit." He recommended the committee schedule hearings with Natives and other residents of the territory before taking any action. "It would seem that enough monopolies have been established in the overexploited territory of Alaska."[2]

The editor of the *Fairbanks Daily News-Miner* was outraged. The bill was necessary to "legitimize" fur farm squatters ("poor people struggling to get a foothold"). He railed at LaGuardia, the Alaska Native Brotherhood ("controlled by William Paul"), and Dan Sutherland, Alaska's nonvoting delegate to Congress. The editor claimed that Sutherland, "dominated by the Indian machine," had failed to speak up to support the bill when "his friend" LaGuardia attacked it.[3] In the end, nothing came of LaGuardia's recommendation for hearings. The bill passed and became law.[4]

The minimum yearly lease fees were set at $25 for mainland farms and $50 for island farms. In addition, lessees were to be charged 1 percent of each year's pelt or breeder sales. Fees higher than the minimum were set by regulations and varied over the years. In 1932, a sample year, the charge for 100 to 500 acres was $50; for 500 to 800 acres was $50 to $100; and for more than 800 acres was $100 to $250.[5] The fur farmers were satisfied, but for Alaska Natives the Alaska Fur Farm Leasing Act of 1926 was an opening skirmish in their broader struggle for land claims. Sutherland would be back in Congress in 1930, introducing an Alaska Native Brotherhood–backed bill that would allow southeast Alaska Natives to sue the United States government for the value of the land they had lost when it was handed over to fur farmers and canneries. The 1930 bill died in committee, but a similar bill finally passed in 1935.[6]

With their land issues settled for the moment, Alaska fur farm associations put energy into getting veterinary help. Farmers were having little luck with home remedies for foxes whose pelts were spotted by mineral deficiencies or mange, and for mink that chewed off their own tails. Moreover, animals were being lost to diseases such as distemper, encephalitis, dysentery, pneumonia, rickets, botulism, hookworms, and lungworms. The fur farm associations pointed out that the territory had no veterinarians and desperately needed one experienced with furbearers.

Basal Parker: Veterinarian and Fur Farmer

Actually, it was not quite true that there were no veterinarians in Alaska. Fur farmers around Afognak Island just north of Kodiak did occasionally get free veterinary services from Basal C. Parker, DVM, who ran a fox-muskrat-beaver farm on Whale Island. Parker had come to Alaska in the summer of 1919 as a Bureau of Biological Survey veterinarian.[7] He was a new graduate assigned to test the territory's cattle for bovine tuberculosis, which can be transmitted to humans. It must have been discouraging work. None of the 650 dairy cows in the territory had previously been tested. Parker identified 103 with positive TB tests and had to destroy them on the spot, leaving dismayed owners with the problem of extracting reimbursement from the government.

In spite of a summer of difficult work, Parker found himself attracted to Alaska. In the fall of 1920, he resigned his government position, brought his wife Madge north from Kansas City, Missouri, and began setting up a

Basal and Madge Parker on their porch of their house at Whale Island. *Baranov Museum, P-639-11.*

The Peak Years 1925–1929

fur farm on Whale Island. This approximately thirty-square-mile island in the strait between Afognak and Kodiak Islands rises to a 2,000-foot rocky peak in the south, but the northern half is made up of lowlands with several small lakes and ponds. Off and on since 1885, the island had been operated as a blue fox farm by various individuals and partnerships, including the Semidi Company. Parker decided that in addition to blue foxes, he would raise muskrat and beaver in the island's lakes.

Parker, outgoing and robust, quickly became well known in the region. Not only was he willing to help farmers with sick animals but also during the winter when boats from the Afognak area could not reach Kodiak, he would occasionally treat an urgent human medical problem with veterinary drugs. His name even entered the local vernacular after a trip he made to Kodiak in his boat, *Prickly Heat*. When he finished buying supplies and was ready to return home, his outboard engine would not start. In front of an interested audience, he sweated to get it going, without success. According to one witness, Parker finally left the harbor, walked to the store and returned with a new sixteen-pound sledge hammer. Back at the boat he took off his coat, rolled up his sleeves, spit on his hands, hefted the hammer and brought it

Crates of live muskrats shipped from Basal Parker's farm on Whale Island to Andrew Grosvold's farm at Sand Point in 1932.
Photo courtesy of Mary Graves Zahn.

down repeatedly on the offending motor. The crowd scattered as metal began to fly. When he had finished the demolition, Parker announced with satisfaction, "There is one engine that won't give anyone cranking problems." From then on, when a piece of machinery was recalcitrant, some local wag was sure to suggest the solution: "Just Doc Parker it."[8]

Parker was not only a man of action, he was knowledgeable about animals. He soon had one of the largest fur farms in the region. His twenty pairs of blue foxes grew to a herd of 200. Moreover, he had begun farming early enough that for many years he was able to sell live muskrats and foxes to farmers who were just getting started. He and Madge built a spacious house on a small bluff with a long porch overlooking the beach. They filled it with domestic amenities such as a short wave radio, an indoor bath, and a billiard table.[9] But although Parker volunteered his services to fellow members of the Southwestern Alaska Blue Fox Farmers' Association "insofar as time taken from [my] ranch activities will permit,"[10] he actually had little time and no laboratory facilities. Farmers in the Kodiak area and elsewhere pushed for the Bureau of Biological Survey or the territorial government to hire a veterinarian dedicated to the industry.

The Bureau of Biological Survey had periodically sent veterinarians like Basal Parker to Alaska on special tasks, most commonly to monitor the reindeer herds that were expected to become a major meat-producing industry. But the Bureau was not interested in sending to Alaska a full-time veterinarian specializing in furbearers. The government biologists believed that an experimental fur farm located within the territory would provide greater help to farmers.

The National Experimental Fur Farm

The Bureau of Biological Survey was enthusiastic about experimental farms and, in 1923, had established the National Experimental Fur Farm at Saratoga Springs, New York. This twenty-acre farm consolidated experiments from the National Zoo and from farms in several states. The first director of the national farm was Frank Ashbrook, fresh from his tour of Alaska. He had just completed an updated, sixty-page bulletin, *Silver Fox Farming* (prepublication sales 10,000 at 15¢ each). With Ernest P. Walker, he would soon co-author another bulletin, *Blue Fox Farming in Alaska*, which would circulate widely in the territory. Ashbrook's publications, although encouraging to farmers, also sounded cautions. He warned new silver fox farmers to start with just a few animals: "Many troubles and obstacles arise, the remedies for which can not yet be found in books, but must be . . . learned through experience."[11] Regarding Alaska blue fox farms, he noted a "number of failures and abandonments of undertakings," due to ranchers leaving island foxes in the hands of incompetent caretakers or simply leaving the animals to fend for themselves for months at a time. "Fur farming requires the same attention and energy that is necessary to success in any other business."[12]

The Biological Survey had initially stocked the Saratoga Springs farm with foxes, martens, badgers, and rabbits. The staff started experiments to evaluate types of food, breeding strategies, and worm medicines. Within two years, the scientists were able to demonstrate that "Samson" foxes lacked guard hairs for genetic reasons and that neither medicines nor diet (tried by Basal Parker and other veterinarians) could cure them. Soon the Survey added mink to their experimental farm and expanded research to include humane killing methods, polygamous fox mating, and vaccines against infectious diseases like distemper.[13] However, the Bureau biologists recognized that Saratoga Springs was distant from Alaska in both miles and climate. They pressed for a territorial experimental fur farm that would investigate disease and nutrition in the north and that would provide scientific data to farmers. But Alaska's fur farmers were looking for more immediate and direct help—specifically, a veterinarian dedicated to the problems of furbearers—so they turned to the territorial government.

Earl Graves: First Territorial Veterinarian for the Fur Farms

In 1925, the newly appointed governor of Alaska was George A. Parks. Unlike many territorial governors, George Parks was an Alaskan.[14] He had come to Alaska in 1907 as a mining engineer. In the course of his work as a federal mineral examiner, cadastral surveyor, and chief of the General Land Office, he had traveled extensively in the territory. His friends included several fur farmers, notably hunting companion Earl Ohmer, who would eventually own the largest mink farm in Alaska. Parks had high expectations for the fur farm industry. He was supportive when, during the 1927 session of the legislature, Representatives Sumner Smith and Ben Grier of southcentral Alaska introduced a bill to employ a territorial veterinarian to aid fur breeders. The veterinarian, working under the governor, was to visit fur farmers, study their problems, and offer advice on care, breeding, disease prevention, and treatment of illness. In addition, the territorial veterinarian was to submit reports to the governor and the legislature on the status of the industry. The bill passed unanimously with a two-year appropriation of $15,000—half of which was for salary and half for travel, equipment, and vaccines.[15]

Governor Parks hoped to hire a veterinarian with furbearer experience from the ranks of the Biological Survey, which had agreed to publish any papers written by Alaska's new territorial veterinarian. But none of the Survey's experienced employees wanted to leave federal service for a job that meant traveling most of the year to remote farms with few amenities and a salary dependent on the munificence of a territorial legislature.[16] Finally, in mid-July the Survey found a willing candidate. Earl Francis Graves, 29 years old, was a recent DVM graduate from Kansas State College with a strong academic record. He had never been to Alaska and had no fur farm experience, but he was athletic, confident, and, like many of the fur farmers, an army veteran. The Biological Survey arranged for him to train for a month at the National Experimental Fur Farm in Saratoga Springs before embarking for Alaska.

Graves reached Juneau in late August and began his first day of work on September 1, 1927, by boarding the *Virginia IV* bound for the Inian Islands and Sitka. In the coming months, he would travel almost constantly by boat, train, and foot with a goal that could not be accomplished in a single year—a personal survey of the fur farms of Alaska. Each week he sent a report to the governor on forms supplied by the Biological Survey. At the end of every month he submitted a narrative summary of his investigations and conclusions based on what he had seen. He was also required to submit an annual report for the governor and territorial legislature.

Graves worked six days a week in conditions that ranged from pleasant to arduous to hair-raising. For example, in July 1929 he enjoyed three nights in the luxury railroad hotel at Curry performing a newly assigned duty—TB tests for the cows that supplied milk for the dining room. The following day he traveled to the whistle stop of Caswell where he walked nine miles to inspect two mink farms. In December he traveled to a fox farm on Eleanor Island in Prince William Sound. He waited for four days to get a boat to

Earl Graves holding orphan fox pups who survived on his recipe for homemade baby fox formula. Photographed at Guy Turnbow's fox farm near Fairbanks, 1929.
Photo courtesy of Mary Graves Zahn.

An unusual Alaska mink farm, probably in the railbelt, where animals ranged freely within a large perimeter fence.
Wennerstrom Collection; Anchorage Museum, B1994.2.233.

the island. Then after working for more than a week alongside the farmer to demonstrate trapping and pelting techniques, he waited three days for weather before attempting to return to Cordova. In retrospect he should have waited one more day. His December 28 daily report reads, "Started to cross Sound in 29 foot boat & storm came up. Were lost as could not see for fog & snow. Spent nite in more or less protected bay." On the bottom margin of the page he comments, "This area is difficult to traverse in wintertime."[17]

Summertime travel was less hazardous, but no less demanding. In July and August 1928, Graves managed to visit farms at Kenai Lake, Russian River, Lawing, Seward, Anchorage, Matanuska, Wasilla, Montana, Curry, Big Delta, Copper Center, Chitina, Cordova, Valdez, and five islands in Prince William Sound. While on the road, the veterinarian often received messages from farmers with animal health crises to which he responded with telegraphed advice. (A telegram on feeding orphaned fox pups began: "FIND NURSING CAT IF POSSIBLE OTHERWISE USE MEDICINE DROPPER BULB FOR NIPPLE. FEED EVERY TWO HOURS." A recipe for fox baby formula created from canned milk, lime tablets, egg whites, and cod liver oil followed.)[18]

Exporting Live Breeders from Alaska

What did Graves conclude from all this travel and observation? He reported to the governor that, in general, mink farmers were doing better than silver fox farmers, who were doing better than blue fox farmers.[19] Mink coats, requiring sixty to eighty pelts each, were beginning to appear in American fashion ads, and the market was responding. In 1925, the average pelt price had been approximately $7. By 1929, it would reach $21.[20]

Not surprisingly, the number of mink farms and the demand for breeders rose with the prices. In 1924, the Bureau of Biological Survey identified in the territory twenty-one mink farms—almost all in southeast Alaska. By 1929, there were at least 153 farmers raising mink, nearly half of them along the Richardson Road or the Alaska Railroad. In that year, customs records showed 729 live mink shipped to the states and to other countries.[21] The new mink farms were located near good transportation, and the farmers had the benefit of two recent publications. *Mink Raising in Alaska* by Gerrit Snider, who farmed near Wasilla just north of Anchorage, was a fifty-page book of

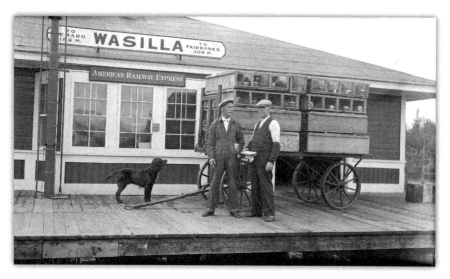

Cart with crates of live mink waiting for the train. Shipper Gerrit Snider (left), author of *Mink Raising in Alaska*, ran a successful fur farm at Lucile Lake. Anchorage Museum, John V. Pulling Collection, B1989.31.19.

advice based on experience: "Don't Buy 'Proven Breeders'! . . . A young mink is more easily raised in a new environment than an old mink is." "Don't Buy Wild Minks! . . . .They will not, except in rare instances, eat bread and milk, and they always remain wild."[22] Frank Ashbrook's slimmer government bulletin, *Mink Raising*, offered scientific advice on sanitation and nutrition that was followed by most Alaskan mink farmers.[23]

Alaska's silver fox farmers, although not doing quite as well as mink farmers, were happy to see silver pelt prices increase from $90 in 1924 to $125 in 1929.[24] Production was high but demand had risen to meet it. National and international fashion was, for the moment, on the farmers' side. In July 1926, a New York reporter wrote that in one hour he observed forty-six women on Fifth Avenue wearing silver fox "scarves," each a single sewn pelt complete with head, paws, and tail. He described the draped style—"over the right shoulder, across the back and around the left side . . . with the [tail] brush hanging down the left side of the back." Another reporter stated that when London passengers debarked from the luxury liner *Leviathan*, "every lady was wearing a silver fox scarf."[25]

The Peak Years 1925–1929

Veterinarian Graves told the governor that silver fox farms of five to ten animals only a generation or two away from the wild were struggling, but that large, well-managed farms were making money, especially those that were shipping breeders to the northern states and other countries. In 1924, for example, Grover Cleary exported from Conclusion Island near Wrangell 300 silver foxes, including eight pairs bound for Germany. And in 1926, Kenneth McCullough of the Alaska Western Fur Corporation accompanied his shipment of 150 silver and blue foxes to Sweden to ensure their safe arrival.[26] Shipments of breeding pairs to the States were so common that the Seattle Box Company advertised their "Fox Shipping Coops" in Alaskan newspapers. The American Express Company took out advertisements listing their rates for transporting live foxes to Seattle from fifty-seven towns in Alaska and the Yukon Territory. The charge per hundred pounds of live fox varied from a low of $6 from Juneau to a high of $34 from Iditarod.[27] In 1929, the U.S. Bureau of Customs recorded 3,333 live furbearers—valued at $156,695—exported from Alaska.[28]

Guy Turnbow, whose farm was on the Richardson Highway about fifty miles southeast of Fairbanks, was one of the better-known farmers who shipped foxes out of Alaska.[29] In 1929, he sold silver fox breeders to Fromm Brothers of Wisconsin, the largest fur farm in the United States. The previous year, Fromm Brothers had sent 8,000 silver pelts to market and received $1.5 million. Henry Fromm, one of the four owner-brothers, traveled to Fairbanks to inspect Turnbow's animals and spoke enthusiastically to a Fairbanks reporter about the local climate, which produced thick fur. His company raised their pups in southern Wisconsin but each fall they trucked those to be pelted 200 miles north to develop better winter coats. The thicker fur added an estimated $50 to the value of each pelt. Fromm was excited about the proposed international road through Canada to Fairbanks, and envisioned Washington state foxes selected for pelting being trucked to Interior Alaska each fall, and "if trucking proved too slow . . . it might be advantageous to use airplanes."[30]

Another large buyer of Alaskan silver foxes was George L. Morrison, who had farmed at Tolovana Hot Springs through World War I. He and his partner, J. Edgar Milligan, had moved their remaining stock to Prince Edward Island in 1920—a fact that made the newspapers because one of the foxes leapt overboard from the SS *Alaska* and swam to shore but could not be found despite a diligent search. In the same year, Milligan and Morrison exhibited live foxes at several Canadian silver fox shows, winning three times as many blue ribbons as their nearest competitor. At a Montreal auction, four of their pelts sold for a total of $8,600. Their reputation secured, they launched a chain of franchise fox farms, to which they supplied breeders as well as advice on management, nutrition, and marketing. By 1924, there were already fifty-two "M&M" franchise farms in Canada and the United States.

That summer, Morrison was back in Alaska to buy 240 silver fox breeders at $500 per pair from Billy and Mickey Williamson in Kasilof and to sell his remaining holdings in Tolovana. Milligan and Morrison were buying animals from other breeders because their Prince Edward Island farm could not produce enough foxes to supply the rising need from what would eventually become eighty-four franchise farms. In November 1926, they set a record for live fox shipments out of Canada when three Canadian National Express railcars filled with 855 silver foxes crossed into the United States to be gradually unloaded at M&M farms along a 5,000 mile route.[31]

Disease, Negligence, and Moonshine in the Islands

During the 1920s, when the popularity of silver fox was increasing internationally, blue fox sales remained centered in the United States. Although Alaska still supplied an important part of this market and shippers sent over 750 live breeders out of the territory each year, island farmers with high transportation costs were beginning to feel competition from pen farmers on the mainland and in the States. Blue fox pelt prices roller-coastered between $45 and $101 during the second half of the 1920s. Sometimes pelts gained or lost 30 percent of their value from one year to the next. In addition to unstable prices, island blue fox farmers were having trouble contending with disease. Many Alaska farmers, as Earl Graves observed, were either pioneers who began before the era of Biological Survey farm bulletins or men who started during the 1920 rush with high expectations and little husbandry experience. Neither group tended to welcome advice from a cheechako animal expert.

Graves believed that Harry Race of Ketchikan probably spoke for quite a few when he complained that "if correct fox farming meant all I had advocated, . . . it was too much work." Graves despaired over owners who did not know how many animals they had or what kind. (Two farmers who complained of no increase in three breeding seasons unwittingly owned only females). He was deeply pained when he discovered on one island some malnourished, parasite-ridden foxes, like "walking, nearly naked skeletons." He was disgusted with men who persisted in feeding their animals deer hair to cure worms and ground glass to cure distemper. And farmers who abandoned their animals for long periods or left them with alcoholic or negligent keepers made him angry.[32]

Alcohol, specifically moonshine, was a problem on island fur farms, not just according to Graves, but according to others acquainted with fur farmers. Alaska had been given a four-year head start on Prohibition when Congress passed the "Bone Dry Law" for the territory in 1916. The law created a shady business opportunity for certain miners and fox farmers who lived beyond easy reach of U.S. Marshals. A Kodiak area old-timer, Ed Opheim, stated that even Long Island—home to Washburn's Semidi farm and Erskine's Kodiak Fur Farm—at one point became "just a front, as so many fox islands were," for the moonshine business.[33] Earl Graves was not a teetotaler. In one report to the governor during Prohibition he praised the hospitality of fur farmers who welcomed him to their table with "meats and fowl, excellent vegetables, [and] splendid jellies. . . . If they have wines or other liquors they are brought out and we talk fox."[34] However, Graves had no patience for men who concentrated on their stills and neglected their animals.

Although Graves found a few expertly managed blue fox farms, others seemed so beset by negligence and ignorance that, as he reported to Governor Parks, "My advice to these men was to kill their animals, sell what equipment they could and go out of the fox game." His frankness provoked angry letters to Parks, who urged his veterinarian to adopt "a conciliatory attitude" and try to "persuade" farmers he was correct, rather than saying that "unless they change their methods, it is useless . . . to visit them."[35] Graves almost gave up after five months on the job, but instead he heeded the governor's advice and settled into Alaska, bringing a bride, Katherine "Kitty" Faulconer, from Kansas to join him on his rounds of the farms. The complaints ceased, and Graves's reports began to mention more fur farmers who, like Peter Wei of Grindall Island or Guy Turnbow near Fairbanks, actively sought veterinary advice and followed it.

Kitty Graves, who traveled extensively with her veterinarian husband, standing near sign required by territorial law. Knight Island, Yakutat, 1928.
Photo courtesy of Mary Graves Zahn.

The Alaska Game Commission and Fur Farm Statistics

At the end of the 1920s, neither Graves nor anyone else was sure how many fur farms were operating in Alaska or how many types of furbearers were being raised. But one government entity tried to find out. The Alaska Game Commission was a joint federal/territorial board that was established by Congress in 1925 to regulate hunting, guiding, trapping, and fur farming in the territory. Four of the five board members were appointed by the governor—one from each judicial district. The fifth, an *ex officio* member, was the chief representative of the Biological Survey in Alaska. For the first few years, this representative was Ernest P. Walker, co-author of *Blue Fox Farming in Alaska* and previously the secretary of the Southeastern Alaska Blue Fox Farmers' Association.

One of the initial acts of the board was to establish a schedule of licenses for hunters, guides, trappers, and fur farmers. The board set $2 as the fee for an annual fur farm license. This license covered one location and one species. Farmers were supposed to buy extra licenses if they used two islands or raised three different kinds of animals. Fur farmers had resisted a $10 license earlier in the decade, but they were more compliant with the $2 license, perhaps because many were also buying the newly created $1 hunting license. Few fur farmers, however, bought more than one farm license even if they were raising several species of furbearers.

Nevertheless, the license lists do give information from the men who were forthcoming about their operations. A. C. Goss was using five islands near Unalaska. Andrew Grosvold—who had gotten an early and well-capitalized start in fox farming at Sand Point thanks to the $100,000 he had earned from selling a Nome gold claim during the rush—was using seven islands. Knute Lind of Minchumina had a single farm on which he was raising silver foxes, mink, marten, lynx, ermine, land otter, and chinchilla rabbits. Don Adler, who played the organ for silent movies in Fairbanks, was raising silver foxes, mink, muskrat, beaver, and chinchilla rabbits on his Richardson Highway farm. In contrast to these men, many farmers never got around to buying even one license—an omission for which there was essentially no penalty. Game wardens, busy with illegal hunters and out-of-season trappers, had no time to check fur farmers' paperwork. Remarkably, only about half the 61 farmers who received propagation permits from the game commission in 1929 actually held a fur farm license. Nonetheless, license data did suggest that the number of fur farms was increasing. In 1926, the new commission granted 262 fur farm licenses; in 1929, it issued 536.

In addition to lists of licenses and propagation permits, other official records of fur farms in 1929 exist—the blue fox branding records, forest service permits, land office records, and an agricultural census of Alaska. In 1929, census takers visited 233 fur farmers scattered through the territory's four judicial districts. They collected data on the number and kinds of furbearers on each farm, how long each farm had been running, its acreage, and the citizenship of the owner or operator. In addition, this census counted some 3,000 furbearers being raised "not on farms"—usually this meant in someone's backyard, although in the case of chinchilla rabbits (some as far north as Kwiguk on the Seward Peninsula), it probably meant inside someone's

Silver fox pups being raised behind a house.
Buzby and Metcalf Photograph Collection, UAF 1963-0071-00056, Archives, University of Alaska Fairbanks.

house.[36] These small "backyard farms" brought extra income to roadhouse owners, trappers, villagers, and miners who had access to fish and game for feed. It also added to the income of a few city dwellers, particularly those in Anchorage where zoning had not yet been introduced and ample city lots were home to large gardens, pets, and income-generating projects.

A summary of the Alaska agricultural census indicates that about one-third of the established fur farmers sampled were unnaturalized immigrants. Over half of these had come from Scandinavia or Finland, another quarter from western Europe and the British Isles, and the remainder from countries scattered across the Northern Hemisphere from Bohemia to Japan.[37] In addition to furbearers, 30 percent of the farms also raised chickens, cows, sheep, goats, or horses. Some 20 percent raised more than one species of fur animal. On the 233 farms there were a total of approximately 10,800 blue foxes, 800 silver foxes, 3,600 mink, 4,000 muskrats, and 5,000 animals from eight other furbearing species.[38]

If one combines census records, licensee lists, propagation permits, general land office records, and blue fox brand records, and then eliminates name duplications, it appears that 622 private farm owners were identified by at least one government agency in 1929.[39] It is difficult to guess how many additional fur farmers operated without coming to government attention. However, maps of Sitka Sound and Kachemak Bay labeled with the names and dates of fur farms as well as the 1929 issues of *The Fairbanks Daily News-Miner* and *The Fur Farmer Magazine* reveal at least forty farmers who were known or newsworthy but not on any government list. (These included Fairbanksan Dr. O.D. Albery, who was unique in raising native marmots for fur.[40]) Considering the widespread practice of backyard farming and the likelihood of other unlisted, unpublicized small-farm operators, the total number of Alaskans raising furbearers in 1929 must have well exceeded seven hundred. Territorial Veterinarian Earl Graves had a large field to survey.

In early 1929, the territorial veterinarian's formal report to the governor and legislature began with this declaration: "It was impractical to visit every ranch in the Territory." But, Graves stated, his travels had made it clear to him that "fox ranching in Alaska has not proven to be as ridiculously profitable as was generally supposed." He had two major recommendations. The first was directed to the territory's island blue fox farmers: Free-ranging foxes on islands were developing a pattern of serious health problems. Farmers needed to take control of their animals. "Control" meant penning foxes for at least part of the year so that pregnant females could receive special rations, young could be fed fish oil to prevent rickets, all animals could be wormed, and specific pairs could be bred. The second recommendation was directed to the governor and the legislature: Alaska should establish an experimental fur farm.[41] As Graves rather plaintively wrote, "Without a single correctly run fox farm in Alaska . . . or a demonstration and experimental plant back of me, I resemble a salesman without a sample case."[42] Moreover, Alaskans had identified special problems that were not being investigated elsewhere—a new species of warble-fly that ruined fox pelts and a mysterious fatal illness in mink that would ultimately prove to be a vitamin deficiency from a steady diet of cannery-salmon waste.

Governor Parks was attentive to Graves's recommendations but judged the cost of establishing an experimental farm to be too high. Instead, he asked lawmakers to pool two appropriations of the previous biennial legislature:

$15,000 for the territorial veterinarian's fur work and $10,000 for the itinerant federal veterinarian's cattle testing work. The resulting $25,000 would be used to create a Department of Animal Husbandry at the Alaska Agricultural College and School of Mines in Fairbanks. The territorial veterinarian would head this department and set up animal experiments at existing, well-run fur farms. In addition, the territorial veterinarian would continue to directly aid fur farmers and would do the required cattle testing. The legislature, however, declined to fund the experimental work. It simply appropriated $22,000 for both dairy and furbearer duties. The governor assigned both tasks to Graves.

As a result of the legislature's decision, Graves's workload increased and an acrimonious struggle for his time erupted between dairy farmers and fur farmers. An attorney general's opinion was required to quiet the newsworthy din.[43] Then the governor, citing the dangers of nepotism, refused a request that Kitty Graves be paid to do her husband's typing. Graves was directed to use professional contract typists. In response, the veterinarian's monthly reports, from that time forward were submitted in his stylish cursive. They also became shorter and shorter. When the governor did not receive an annual report in January 1930, he pressed Graves and in April issued a forty-day ultimatum.

Graves responded with a letter of resignation: "I desire to enter into the commercial production of fur . . . I will now have an opportunity to demonstrate the facts I have been advocating and to develop a few theories."[44] Graves's opportunity was a fur farm in Spokane Bridge, Washington, but his timing was inopportune. The stock market had crashed in October 1929, and the effect on pelt prices would be felt in the winter fur markets of 1930–1931.

The phrase *The Great Depression* was not yet in use, but the market for both luxury and utilitarian furs was poised to nosedive. Earl Graves and hundreds of fur farmers who remained in the territory were about to join fellow citizens throughout the country in facing hard times.

CHAPTER SIX

The Great Depression 1930–1940

Depression Years and Alaska's Experimental Fur Farm

As the Great Depression settled over the United States, smothering the vibrant fur markets of the 1920s, Alaska fur farmers watched their income tumble. Blue fox pelts that had sold for $101 in 1929 fell to $21 by 1932. Mink slid from $21 to $6, and silver fox toppled from $125 to $41.[1] At first, many fur farmers, their associations, and their journals clung to the optimism of the boom years. Markets had fluctuated in the past. Alaska fur farmers reminded each other that Joe Ibach and others who had hung on when prices crashed at the onset of World War I had done well in the end.

Initially, the U.S. Biological Survey was also optimistic. The Survey mounted an exhibit, *Fur Resources of the United States*, for the 1930 International Fur-Trade Exposition in Leipzig, Germany. America's fur farms were the centerpiece. The fifty-page brochure for the exhibit was written by Frank Ashbrook, now head of the Biological Survey's Division of Fur Resources. He estimated the U.S. investment in fur farms at $50 million and noted that 98 percent of American silver fox pelts came from farms. Photo-filled sections of the booklet described farms successfully raising silver fox, muskrats, mink, raccoons, beaver, and rabbits. The final third of the brochure was dedicated to the U.S. government's blue fox farm and fur seal harvest on the

Pribilof Islands. The Foreword concluded, "Fur farming . . . is surviving in very healthy condition . . . and now is established on a firm basis."[2]

In 1930, Alaska communities also staged fairs that touted local fur farms. Nome held its first fur breeders fair that year. Growers displayed white foxes native to northwest Alaska as well as silvers and blues transplanted from farms elsewhere in the territory. In Anchorage, the Cook Inlet and the Susitna Mink Breeders Association had just formed. The Anchorage Agricultural Fair, which had focused on silver foxes in previous years, now also featured mink. In addition, the Anchorage fair offered cash prizes for marten, muskrats, and blue foxes. Contestants entered more than a hundred animals. The Seventh Annual Tanana Valley Fair at Fairbanks, which claimed that there were more than two hundred fur farmers in their vicinity, issued awards for chinchilla rabbits as well as for larger furbearers.

Articles in *The Fur Farmer Magazine* described the fairs of 1930 and 1931 and tried to remain upbeat. In October 1931, the editor printed an encouraging bulletin from the Seattle Fur Exchange:

> In view of the fact there is no surplus of Raw Furs in any market, we believe the demand will continue to be steady to strong, at present price levels. During this entire upset economic condition the fur trade has been very fortunate inasmuch as they have been able to dispose of their merchandise at fair value while other raw commodities have suffered intensely. . . . The scarcity of desirable Raw Furs is growing more acute.[3]

Increased efficiency, cooperative marketing, and better fashion advertising were proposed to alleviate the economic downturn, but fur farmers everywhere found their situation increasingly dismal. Sale of breeders was at a standstill. In Alaska in 1932, customs officials reported no mink and only two blue foxes shipped to buyers outside the territory, a far cry from the more than 3,300 animals shipped in 1929. The pioneering Sholin brothers of Homer, whose main business between 1918 and 1929 had been the sale of silver fox breeders at $600 to $700 per pair, went out of business in 1934.[4] Pelts—the farmers' only remaining source of revenue—were steadily losing value. As for reassurance from the Seattle Fur Exchange, farmers already knew that the owners of the exchange had badly misjudged the market when they expanded into live breeder sales in 1930.

The Great Depression 1930–1940

Anchorage Agricultural Fair prize winner from Williamson Silver Black Fox Ranch.
Photo courtesy of Mary Graves Zahn.

Articles in the 1931 issues of *The Fur Farmer Magazine* were cautious. They predicted a decrease in both demand and price during the near future. Nevertheless, the magazine writers projected a strong, long-term recovery for the industry. One noted that the price decreases of 1930–1931 had especially caught silver fox farmers by surprise because silver fox, a premium fur, was expected to hold its value better than blue fox or mink. But silver fox lost a quarter of its value in just two years (and was to lose two-thirds in the next two years). Farmers who had reinvested profits into increasing their breed stock rather than sending pelts to market were low on cash. Many would be forced "to wind up their fur breeding ventures next winter and pelt all their animals. . . . The high cost ranch will not be able to exist." However, the author asserted, "The fur market situation will right itself."[5]

But two years into the Depression both farmers and the Biological Survey were beginning to question the market's resilience. In the January 1932 issue of *The Fur Farmer Magazine*, Paul Redington, head of the Biological Survey, warned that in a depression, furs are "the first to register a decline and generally the last to recuperate." He recommended that farmers cut costs and diversify by adding other crops and perhaps by raising several species of furbearers because it was no longer clear which furs would bring the best return in a volatile market.[6]

Despite this growing pessimism, the editor of *The Fur Farmer Magazine* was determined to remain optimistic. In March 1932, he wrote:

> There never was a better time than the present to begin fur farming. . . . The well known depression we hear so much about will pass. It is hurting too many important people to be allowed to continue. . . . When the depression passes the world will require furs. . . . The necessary quantities are not in the woods. Nor are they in warehouses waiting. . . . But the fur farms can provide them. If we can only see a little past the blue present. . . . No market is more insatiable than the fur market when it gets its stride. No market is more generous in the matter of price.[7]

Closing Farms and Changing Fashions

Optimism aside, fur farms were closing, not starting up. In Alaska, feed and transportation costs were high, and diversification to other furbearers or other crops was seldom feasible. In 1936, the Alaska Game Commission issued only 243 fur farm licenses and 7 propagation permits—down from 536 licenses and 150 permits in 1929.[8] Several fur farm associations in the territory dropped their newsletters and ceased meeting, as did many of their stateside counterparts. The long-established National Association of the Fur Industry liquidated. Sales at the Seattle Fur Exchange continued to slump. *The Fur Farmer Magazine* lost subscribers and advertisers. Classified ads that used to occupy two columns shrank to half a column. Notwithstanding its strenuous efforts to retain readers, the magazine was forced to switch from glossy pages to pulp paper in December 1934. The next month, it turned into a weekly news sheet. Six issues later it went out of print.[9] Across the country in New York, *The Black Fox Magazine*, a trade journal dating back to World War I, also folded.

Women's fashions changed with the onset of the Depression. Many designers abandoned the fur market. Only Hollywood costume shops and a coterie of still-wealthy women were buying luxury furs. Gone were the days of flappers with long fur coats covering bare knees in open cars. When the transatlantic liner *Queen Mary* arrived in New York after her maiden voyage in 1936, few of the elegant women who descended the gangplank had new silver fox stoles curled around their shoulders. Hemlines had dropped, and on most of the new long coats fur had been reduced to trim. During the Depression, some outdoor photos of Eleanor Roosevelt showed her in a white fur jacket or silver fox stole, but more often she was pictured wearing a cloth coat, dressing in solidarity with the majority of American women.

Government Aid to Failing Farms

Fur farms in the United States were in trouble, and the government sought to help. In 1930, Congress enacted broad protectionist legislation, including a 50 percent *ad valorem* tariff on silver fox imports that was designed to blunt competition from Canadian fur farmers. Then in November 1933, at the urging of the Biological Survey, the National Reconstruction Administration for the first time classified fur farming as an agricultural enterprise. This allowed the Farm Credit Administration to issue short-term loans to help fur farmers get through the traditionally difficult period of October through December. During these final months of the year, animals needed supplemental food to develop thick coats. Often scant cash remained from the sale of pelts during the previous January. Widespread bank failures made commercial loans to small borrowers virtually unavailable, but after 1933, fur farmers could turn to the Farm Credit Union. Finally, as the Depression began to improve in 1936, the federal government sought to aid the industry by stimulating sales to consumers. The federal excise "luxury" tax on furs was reduced from 10 percent to 3 percent. Two years later, the tax was abolished altogether.

In Alaska during the first three years of the Depression the Forest Service decreased its fees for fur farm permits—initially by 20 percent and then by 50 percent. Nonetheless, many fur farmers sent the government IOUs or simply failed to pay. By 1935, the Forest Service in Alaska was holding a dozen promissory notes from farmers. Rents had been dropped so low that the average IOU was for only $85, but $85 was more than these permittees could afford. Another 120 farmers had simply made no response to government requests for payment.[10] Local land managers were reluctant to evict farmers and seize improvements for which there were no buyers. As one explained in a memo to Washington:

> I know most of the signers [of promissory notes] and am . . . acquainted with their circumstances. Take Liljegren & Beyer and Mrs. J. B. Clock. They are operating big farms in Prince William Sound. Neither island has paid expenses for years and it has been necessary for occupants to do commercial fishing etc. in order to make a living. I doubt if they ever can pay the $125/yr rental. On the other hand it would be inhuman to make them vacate the islands. Mrs. Clock and Mrs. Beyer were born on their respective islands and have never known any other house. Each inherited the home and business from her father and each has substantial permanent improvements . . . representing all her worldly possessions. To force them from their homes would be to deprive them of all they own. . . . Perhaps [the Forest Service should] grant the occupants a $5/yr. homesite permit . . . for a specified number of years.[11]

The Forest Service adopted this suggestion and began offering failing permit-holders in the Tongass and Chugach Forests the option of converting their farms to five-acre, $5-per-year homesites. But elsewhere in Alaska, the fur farmers who were leasing land from the General Land Office had no such option. Delegates Wickersham (1932) and Dimond (1933) introduced bills that would allow fur farms leased from the Land Office to be converted to homesteads that could be "proven up" and deeded to the claimant, but neither bill became law.

Jule Loftus: Last Territorial Veterinarian for the Fur Farms

Within Alaska's territorial government, support for fur farmers continued to be strong. Governor Parks hastened to fill the veterinarian position that had been vacated by Earl Graves in May 1930. The governor, again unable to find a candidate through the Biological Survey, selected a veterinarian recommended by Charles Bunnell, president of the Alaska Agricultural College and School of Mines at Fairbanks. Jule B. Loftus, one of the first students to enroll at the college in Fairbanks, had completed his veterinary degree in Colorado, where he had done extra work with furbearers. Like Graves, he was a World War I veteran with a good academic record. Moreover, he was not a cheechako, having previously labored on the construction of the Alaska Railroad, hunted sheep for roadhouses, and corralled caribou with biologist Olaus Murie.[12] Parks offered the job to Loftus, but his letter contained a caution: "We are dependent on appropriations from the Legislature and from Congress . . . and cannot guarantee employment beyond the end of the biennium, March 1931."[13]

Loftus, undeterred, accepted the position. He spent a month at the national experimental fur farm at Saratoga Springs, New York. He then visited the two largest fur farms in North America: Fromm Brothers in Wisconsin and Milligan & Morrison on Prince Edward Island. Primed with the latest information on fur farming, pelt grading, and veterinary care, he arrived in Juneau to start work on August 1, 1930.

Like Graves, Loftus used boats, roads, and the railroad to reach farmers, but he was able to travel farther afield than his predecessor because airplanes were coming into common use. A number of fur farmers in distant parts of Alaska, who had never had a veterinary visit, were eager for his advice. His first airplane tour was something of an adventure.

In August 1932, Loftus arranged a flight from Fairbanks to inspect two fur farms in northwest Alaska—trader George Goshaw's at Shishmaref and Hugo Eckardt's on the Noatak River. The plane was a single-engine Bellanca on floats piloted by Victor Ross. Loftus and Ross spent the night at Nome, where Loftus tested Nome's three cows for tuberculosis. In the morning, the men headed across the Seward Peninsula toward Deering but lost visibility in an unseasonal snowstorm and wandered off course in the vicinity of Iron Mountain where the compass was unreliable. Eventually, they spotted a river flowing north into Kotzebue Sound. They landed at its mouth, and spent the night sleeping upright in the plane's narrow seats. The morning proved clear, but Ross decided he was too low on fuel to fly so they taxied along the coast to Deering, a journey of four hours. After refueling, they flew to the Noatak River where trader Eckardt and his wife, a charming former opera

In front of the Bellanca plane at Shishmaref, 1932, L to R: Victor Ross (pilot), Alvin Polet (Nome businessman), George Goshaw (fur farmer and trader), two unidentified Eskimo men, and Jule Loftus (veterinarian).
Photo courtesy of Bruce McNaughton.

singer from Budapest, ran a well-kept fox farm. When the plane landed, a rock punctured a hole in one of the Bellanca's floats. Eckardt conferred with Loftus about adding mink to the silver and arctic foxes on his farm while Ross performed a tar-and-canvas repair. Then they flew to Shishmaref, where Loftus toured Goshaw's outdoor pens, which housed some two hundred foxes. The two men discussed Goshaw's experiments in crossing blue with white arctic foxes, with the objective of producing foxes with pale blue fur that could compete with silver fox. The next day when Loftus and Ross attempted to return to Nome, their airplane troubles continued. They had to turn back twice—once because of falling oil pressure (resolved by purchasing oil from Goshaw) and once because of fog. According to Loftus, Ross quit flying after that trip and became a Deputy Marshal at Sitka.[14]

In his monthly reports Loftus, experienced in dealing with maverick Alaskans, was less dismayed than Graves by wrongheaded husbandry and more often admiring of farmers' ingenuity. For example, he had high praise

Territorial Veterinarian Jule B. Loftus, the second to serve in that position, picking beach strawberries near Juneau.
Photo courtesy of Jule H. Loftus.

for the efficiency and cleverness of Bill Abbes, who raised blue foxes on tiny Storm Island, 80 miles south of Juneau and less than a half square mile in size. In 1931, Abbes was able to raise 203 pups to maturity by running a gasoline-powered cart on 2 × 4 wood rails to feeding stations and fox dens. None of his pups had to venture more than 6 feet from a den to reach food and water.[15]

Overall, the reports that Loftus sent to Parks echoed Graves: "Throughout this district the health of the mink leaves nothing to be wished for. . . . "[16] In contrast, free-range fox farms were in trouble. The warm, wet summers of the Tongass and Chugach Forests—quite unlike summers in the foxes' native arctic habitat—allowed hookworms to establish themselves in the soil, and many young foxes died before maturity. Loftus found that "enough [island foxes] survived so that ranchers made money while there was a heavy demand for fur but when prices were poor it was not a profitable business."[17] However, Loftus encouraged farmers to persevere, to pen their animals and improve their husbandry because he expected the economy to improve. In his first annual report to the governor, he wrote, "The prices of raw fur at present do not justify the production on ranches of some of our fur bearers

George Goshaw's fur farm in winter, 1927. The tower overlooks his extensive pens that are faintly visible in the distance.
Alaska State Library, Alaska Road Commission Photograph Collection, P-61-008-089.

but with a return to near normal business . . . the number of fur farmers will increase and the cost of producing pelts will decrease through better methods and increased production."[18]

Fur Farming in the Depth of the Depression

Pelt prices reached a Depression nadir in 1933. The financial damage of the previous three years was insurmountable for most farmers. In 1934, pelt prices began a slow ascent, but Alaska fur farms would continue to close for several more years. Feed costs were partly to blame. Farmers along the railbelt (the Alaska Railroad corridor connecting Seward, Anchorage, and Fairbanks) customarily cooked up rations of wild meat, cereal, and vegetables in huge iron pots over outdoor fireboxes. These farmers were at a special disadvantage in the period from 1930 to 1932 because snowshoe hares in the region were at a low point in their approximate ten-year cycle. Interior fur farmers could not simply replace hares with moose or caribou because territorial law banned feeding game (with the exception of predatory birds) to captive foxes or other furbearers.[19] This law, on the books since 1920, was not vigorously enforced, but the usual penalty—a $100 to $250 fine and forfeiture of the guilty person's rifle—was a deterrent.

A few farmers managed to keep their furbearers fed by killing wild nongame animals. Mink farmer Oscar Christensen, for example, was a heavy user of porcupine meat. He hunted at night and estimated his take at 900 per year—15,000 during the seventeen years he farmed at Moose Pass near the center of the Kenai Peninsula.[20] But when snowshoe hares were in short supply, most railbelt farmers relied heavily on salted fish brought by train from the coast or on the newly marketed mink and fox pellets shipped at prohibitive cost from the States. Anchorage mink breeders banded together to buy pellets in large lots and shared cold storage space for fish. The Anchorage area fox breeders also tried to cut food costs for members of their association by buying cereal in bulk with the result that many foxes in the area received sufficient calories but were deficient in protein. Moreover, farmers using dried or salted fish that contained little vitamin-D–rich fish oil also found their fox pups prone to rickets.[21]

It was easier for coastal fur farmers to feed their stock because they had access to fish, to cannery waste,[22] and to marine mammals—seals, porpoises, and sea lions—which were not considered game. Since 1927, fur farmers had even been able to collect a $2 territorial bounty on each seal they killed for fox food because seals were accused of eating too much of the valuable salmon harvest.[23] Larger sea mammals were also legal prey. For a few years in the late 1920s and early 1930s, Louis Lane, captain of the whaling vessel *Gunner*, had a business delivering whale meat and blubber to island farmers in the vicinity of Kodiak, Cook Inlet, and as far south as Juneau. Tallak Ollestad, a Kachemak Bay farmer, recalled that the price for having a whale killed and beached on his island in the late 1920s was $100.[24]

But despite this bounty from the sea, coastal farmers were struggling. In 1933, many of them lost supplemental income when Prohibition and the lucrative sale of moonshine ended. Moreover, 1933 was the second-worst commercial fishery year in Alaska since the First World War.[25] In the Fox Islands, Aleut fur farmers on Umnak Island had a disastrous year. During a January gale, the 65-foot schooner *Umnak Native*, owned by the village fur farm co-op and carrying trappers and furs, lost power and was slammed onto the rocks. The ship and cargo were a total loss, and eleven of the fifteen people aboard died in one of Alaska's worst shipwrecks of the decade.

Porpoises killed for fox food at a farm on Sullivan Island north of Juneau, 1922.
Photo courtesy of Mary Graves Zahn.

Umnak's fox farm business was also hit hard by the financial failure of a joint venture with a farm in the Andreanof Islands, the Kanaga Ranching Company. In 1926, Umnak had signed a five-year agreement with Harold Bowman, who operated Kanaga Ranching. In exchange for half the village's fur farm permit for neighboring Tanaga Island, Bowman agreed to add five pairs of blue foxes to the island each year, erect several buildings, hire local Aleuts at fair wages, and provide free boat travel across the 400 miles between Tanaga and Umnak. At the end of five years, the net profit would be split between Kanaga Ranching and the Umnak village co-op.[26] Tanaga Island was productive. When the five-year mark was reached in 1931, Kanaga Ranching Company sent 2,145 blue fox pelts from Tanaga and two smaller islands to the Seattle Fur Exchange. A photo of the spectacular shipment was featured on the August cover of *The Fur Farmer Magazine*.[27] In that year, 1931, Kanaga Ranching sold almost six times as many blue fox pelts as the U.S. government did from their operation in the Pribilofs.[28]

One reason for the large difference in sales between the private company and the government was that Pribilof agents had decided to hold back their 1931 harvest because of low prices. Instead of their usual 850 to 1,250 pelts, they shipped only 370 to market. Some of Alaska's private farmers also held back that year's harvest, but it was a poor decision because prices in 1932 and 1933 were even worse. Although Kanaga Ranching had chosen the better strategy, it was not sufficient to make a profit for them or for the Umnak village co-operative. Their large lot of furs, if sold at average prices in 1929, would have brought about $217,000. In 1931, it was worth only $58,000. At the end of the year, Kanaga Ranching owed nine firms (including a bank, a wholesale grocer, a shipbuilder, and a shoe retailer) worth approximately $142,000. The company sold its largest ship and mortgaged all of its assets to a consortium of its nine creditors. The assets included foxes, fur farm permits, pelting equipment, boats, a 300-foot dock, the Kanaga store, and a Seattle office. The monthly payments were to consist of all Kanaga Ranching proceeds less expenses. The combined salary of Harold Bowman and his secretary was capped at $600 per month.[29]

Unlike the Umnak village fur co-op, Bowman's company managed to stay afloat through the Depression. Bowman held onto his Kanaga mansion "complete with Japanese cook and maid," but he had to give up his dream of starting an annual Aleutian Fur Rendezvous.[30] He persisted in fur farming and even diversified in 1936 by importing eleven Karakul sheep. The sheep were intended as the nucleus of a herd that would produce Persian lamb, a fur that was rapidly gaining popularity.[31] Although there is no record of sales from this venture, Kanaga Ranching survived until 1941.

The Alaska Experimental Fur Farm

By 1936, the Alaska fur farmers who remained in business began to feel more optimistic. Fur prices inched upward, and in the territory as a whole the economy was improving. President Franklin Roosevelt had raised the price of gold from $20.67 to $35 per troy ounce. More mines became active, and Alaska reaped a mineral tax. Territorial revenue grew. The Civilian Conservation Corps and Public Works Administration were putting Alaskans to work on public projects. Additional federal money arrived in the territory to fund the Matanuska Colony Project, a program to transplant hard-hit farmers from Michigan, Minnesota, and Wisconsin to a fertile valley north of Anchorage where they could make a new start. *The Fur Trade Journal* responded to news of this program with a long article predicting failure unless the colonists forgot about dairy cows and hay and shifted to a crop more suited to the climate—namely, furs. Ironically, a few years later, the *Fairbanks Daily News-Miner* reported that one of the colonists, Ballard Dean, actually started to farm mink and sold 100 pelts, but then "Colony Director [Ross] Sheely barred mink raising." Dean quit the colony and took his five children south, in the hope of starting a mink farm in Washington state.[32]

Despite territorial budget cuts in the early thirties, veterinarian Loftus held on to his job. He was still employed, thanks to Charles Bunnell, who had arranged for the federal government to pay a portion of the territorial veterinarian's salary through the Cooperative Extension Service. When the economy picked up in the mid-thirties, Loftus, Bunnell, and the Biological Survey all began to realize that an Alaska experimental fur farm was within reach. Ironically, the experiment station that could not be funded during the high economic times of the 1920s looked affordable during the waning years of the Depression. The federal government could donate land, the Civilian Conservation Corps and Public Works Administration could clear trees and erect pens and buildings, and the territorial veterinarian could manage the farm under the Cooperative Extension service.

Loftus had obtained permission from the 1931 legislature to conduct fur experiments on private farms. He had begun by trying to improve fur quality at the Yukon Fur Farm, the territory's largest minkery. Located near Petersburg, the farm was owned by Earl Ohmer and partners. Ohmer, "the Shrimp King of Alaska," also owned Alaska Glacier Seafood Company.[33] Loftus designed an elevated mink pen with wire flooring to house 250 males during the three months they were developing winter coats. To prevent battle scars, the minks' canine teeth were clipped. Feeding was easier, survival was a highly unusual one hundred percent, and the fur was thick and unmatted.

Ohmer was gratified by these results. He was well-known beyond Petersburg. Easy to recognize—six-and-a-half feet tall, pointed beard, customarily clothed in riding breeches with puttees, and sporting a gold watch chain—he was a man of strong opinions and considerable public service. In 1935, after serving Petersburg as councilman, mayor, and Chamber of Commerce president, he was appointed chairman of the Alaska Game Commission, a position that he was to hold for the remaining twenty years of his life. After the success of the mink experiment at his farm, he became a vocal proponent of establishing an Alaska fur experiment station. In 1936, he sent mimeographed petitions to fur farmers throughout the territory, urging them to sign and return the forms so that he could present them in a stack to the legislature.[34] Other well-known Alaskans such as George Goshaw at Shishmaref; Gerrit Snider, author of *Mink Raising in Alaska*; and Wellman Holbrook, regional forester for the Chugach National Forest, joined in the call.[35]

In 1937, the legislature responded by appropriating $20,000 to establish an experimental fur station near Petersburg on land to be selected by a committee of three—Governor John Troy, B. Frank Heintzelman from the Forest Service (which contributed thirty-five acres of land), and Frank Dufresne of the Biological Survey (which granted $4,000 for research equipment).[36] Railbelt farmers had hoped the experimental fur station would be located in Matanuska, arguing that money would be saved if the territorial veterinarian were headquartered in prime cattle country because tuberculosis testing was still part of his job. But Earl Ohmer had prevailed, convincing legislators that the farm needed to be located in his fish-rich area.

The site chosen by the committee was cleared of trees by the Civilian Conservation Corps. The Public Works Administration awarded a building grant and oversaw the building contractor. The new farm was designated

Earl Ohmer, fur farmer, fish processor, and long-time chairman of the Alaska Game Commission.
Photo courtesy U.S. Fish & Wildlife Service, National Digital Library.

The Great Depression 1930–1940

Substation 2 of the Agricultural Experiment Station at the University of Alaska, the new name for the institution headed by Charles Bunnell. Loftus was to divide his time between two jobs—one month each year to veterinarian tasks not related to fur farming and eleven months to superintending the Petersburg station and visiting fur farmers in the field.

Work began on the Petersburg station in the spring of 1937. The contractor erected a two-story, saltbox-style central building with a laboratory, library, offices, and a three-bedroom apartment for the superintendent and his family. Loftus and an assistant constructed a freezer building to store fish. The freezer was an important innovation for fur farming because frozen fish did not have to be salted or cooked before feeding. (One of the earliest studies at the farm was designed to show doubting farmers that coarsely ground, long-frozen fish was safe for animals.) The territorial government purchased $9,700 worth of breed stock from various Alaskan fur farms—30 blue fox, 40 mink, 4 silver fox, and 4 marten. The animals were housed in

Alaska Experimental Fur Farm, 1938. Superintendent's home and office built by the Public Works Administration. Animal cages visible in background.
Agricultural Experiment Stations Collection, 1968-0069-0000b, Archives, University of Alaska Fairbanks.

wire-bottomed pens that lined the inside of six long buildings designed to handle 200 animals. In the rafters of the buildings, the superintendent and his assistant built cages for the white rats and pigeons that were to supplement the furbearers' diet of flounder, halibut heads, and dog salmon.[37]

Animal experiments—many of which would stretch over several years—began in 1938. One early investigation revealed that an all-salmon diet was detrimental to mink. Another measured the survival time of parasites in southeast Alaska soil. Yet another revealed the cause of an inflammatory blindness in fox pups. About $1,000 worth of furs were sold each year to help defray the farm's expenses.

Territorial Fur Farms that Survived the Depression

With the founding of the Alaska Experimental Fur Farm and the easing of the Depression, territorial fur farmers looked forward to the return of prosperity. Pelt prices were still lower than in 1929, but in Alaska both the price and the number of blue fox and mink pelts sold were well above the lows of 1933. The price of each had risen by $5.50. Moreover, blue fox sales had doubled to 11,500 pelts, and mink sales had increased by 40 percent to 40,000 pelts. However, in 1938, the news for Alaska's silver fox farmers was still not good. Although the volume sold (1,100) was back in the 1929 range, silver fox prices continued to drift downward. In 1939, the value of an Alaskan silver fox pelt was almost $15 below its 1933, depth-of-Depression price.[38]

Commentators suggested that silver fox failed to rebound simply because North American and Scandinavian farms had saturated the market. In the United States, silver pelt production had soared from 6,000 in 1923 to 170,000 in 1935. And the United States was not the world leader. In 1936, Norway's production of 230,000 pelts exceeded the combined output of the United States and Canada. Moreover, Norway and her European neighbors had been able to keep production costs down. In spite of the 50 percent import tariff, European farmers found it economical to send thousands of silver fox pelts to U.S. auction houses each year.[39]

At the end of the 1930s, despite the disappointing returns from silver fox, Alaskan fur farmers still in business were generally optimistic. Innovative breeders continued to work with new species. Neva Rusich on Cliff Island near Kake was raising fitch, a mid-sized member of the weasel family native to Europe. Nelson McCrary in Copper Center was breeding nutria, a South American rodent somewhat larger than a muskrat. A. W. Bennett on Long Island near Kodiak was raising black raccoons as well as foxes and muskrats. This latter farm prompted the Biological Survey to send a letter of concern to the Alaska Game Commission about the raccoons, who were escape artists capable of competing with local fauna.[40] Although the earliest recorded Alaskan raccoon farmer, C. E. Zimmerman, had managed to keep animals contained on Brothers Island, other farmers had been less successful. For a decade on Warren Island near Prince of Wales Island there had been a renegade colony of raccoons established by escapees who swam to the island from farms twelve miles to the west. The skill of raccoons in making use of new surroundings was later recounted in a letter to the Fish and Wildlife service by fur farmer Louis Scott. Scott reported that he had placed eight black raccoons on Singa Island near Prince of Wales. Soon there were raccoons on four islands in the vicinity, including El Capitan, where Scott resided and raised blue foxes. He described the situation:

> I have Black Raccoons in my chicken house, in my feed room and in my hair. If I leave any fish or seal meat down on the float at night the little bums pack it off by the arm loads. . . . I have never seen anything as fat as those animals. . . . They are on the beaches rolling over rocks getting crabs, eels, small bull heads, etc., then . . . they cram themselves full of red berries, salmon berries, salal berries etc. How they keep from busting a gut is more than I can see.[41]

The Biological Survey's 1932 letter to the Alaska Game Commission regarding raccoons seems to be its earliest official statement about the potential danger of raising invasive furbearers on Alaska's farms. (However, Olaus Murie would soon complete an eye-opening trip to the Aleutians and raise a major alarm about the environmental risk from foxes.) As for the raccoons, the Game Commission asked the territorial veterinarian to investigate, but after receiving his report, took no further action. The commission continued to issue licenses to farmers of both native and non-native species.

At the end of the Depression, who were the farmers that still held licenses issued by the Alaska Game Commission?[42] Of the pioneers mentioned in previous chapters, the majority had quit, moved, or died before December 1939.

Basal Parker, the veterinarian-farmer, died in a fur farm accident. One January day in 1936, he heard gunshots on Whale Island that he suspected came from poachers. He and a hired helper took a shortcut across the ice of his muskrat ponds to accost them. Parker had grown rotund; the ice broke under his weight, and his helper was unable to pull him from the water in time to save his life. He was 43 years old. The Depression had taken a toll on his holdings. His wife Madge inherited boats, buildings, and foxes valued at only $945. She moved south. In the 1940s, the Whale Island permit was taken up by Bob Van Scheel from neighboring Afognak Island.[43]

George Morrison, who started in Tolovana and co-founded the M&M chain of farms, also died accidentally. He and his partner, J. Edgar Milligan, were on their way to a training conference for their franchise fox farmers in September 1933 when they were killed in an automobile accident. Despite the loss of its founders, the M&M system survived the Depression under new managers.[44]

John Hegness, who operated the white fox farm near Barrow, left Alaska in 1933 to work for MGM Films in Hollywood. In 1925, he had mushed his dogs south to catch a ship and bring footage of the Iditarod serum run to the film studio. He worked for the studio again in the winter of 1932 managing "the MGM polar bear camp" for the shooting of Peter Freuchen's film, *Eskimo*. Hegness continued to work in Hollywood for more than a decade and never returned to Alaska to live.[45]

William Abbes, admired by Jule Loftus for his efficient farming, also moved out of Alaska. Abbes owned two permits—one for Storm Island, the other, acquired in 1933, for Midway Island. In 1935, he dropped his fur farm licenses and moved to Seattle so that his daughter could attend a regular high school. He advertised his farms for sale. He found no outright buyers so he kept the Forest Service permits and arranged for managers who would buy the improvements over time. His brother-in-law took over Midway Island, but was unable to make enough money to buy in. Abbes blamed market conditions and fisheries regulations "that prohibited taking a few fish for cheap feed." The brother-in-law complained to the Forest Service about suffering under an absentee landlord. The Forest Service was already concerned about "sharecroppers" who had no investment in either the fur farm permit or island improvements. The Forest Service wrote to Abbes and advised him to renegotiate realistic purchase arrangements if he wanted to retain his permits. Abbes replied with the fruitless complaint of many fur farmers, stating that after "seventeen years of pioneering and developing" Storm Island, he should be allowed title to the land. When World War II arrived, the two farms were abandoned by their operators, and the Forest Service revoked the permits.[46]

Nicholas Bolshanin was still listed as a board member of the Aleutian Islands Fur Farmers Association in 1940, but he dropped his fur farm license in 1933. After that year, he probably no longer made an annual trip from his residence in Sitka to oversee the foxes on Kavalga and Ulak Islands.

C. E. Zimmerman, who had added raccoons to his island in 1919 and who had written *The Pathfinder* article about his favorite foxes—Rags, Trailer, and Scratcher—was in poor health by the time the Depression was half over. In 1938, he tried but failed to sell his improvements and animals on Brothers Island. After a caretaker pelted most of the foxes and left the island without a farewell, the Forest Service offered Zimmerman a five-acre homesite permit. Zimmerman declined, and the fur farm permit was withdrawn in 1941.[47]

In 1929, Billy and Mickey Williamson, who had developed a successful silver fox farm on the Kenai Peninsula, decided to change locations in order to cut feed costs. They planned to build a new farm at 23 Mile on the Alaska Railroad, where shipping charges for boneless horsemeat were reasonable. The area was also rich in easily hunted porcupines. They expected the move to take two years, but by 1931, they abandoned the effort. Their planned one hundred 65 × 12-foot pens were never constructed. Soon thereafter, the Williamsons went out of business.[48] Others who gave up during the Depression included Don Adler and Guy Turnbow of Fairbanks and Knute Lind of Minchumina.

W. J. Erskine and his partners near Kodiak showed an inspired sense of timing about when to drop out of the fox business. In 1929, they sold their Long Island farm to Rupert Sholtz of the Alaskan Fur Corporation for $14,750.[49] Sholtz hired A. W. Bennett to manage and expand the farm. By October 1929, Bennett had added twenty mink and a sawmill. He bought modern ditch-digging tractors to create muskrat ponds and imported one thousand muskrats from Montana. The muskrats were successfully shipped to Long Island, ten to the crate. From a Wisconsin farm he purchased twenty raccoons, whose descendents would eventually raise concerns in the Biological Survey. In Alaska, Bennett also purchased dozens of blue foxes from local farmers, bringing the herd on the island to more than two hundred animals. However, in less than seven years the corporation was overwhelmed by Depression economics. The lease lapsed, and fifty years of fur farming on Long Island ended. When the United States entered World War II, the military took over the island and transformed it into Fort Tidball.

In contrast, Joe and Muz Ibach were able to hold on to their Lemesurier Island home, even though the Depression essentially ended their fox farming. In 1936, Joe told the Forest Service that he still maintained thirty pairs of breeders and was selling about 120 pelts per year. But he added that prices were very low and poaching was increasing. The following year, he stated, "The price of fox pelts is so low that I have to trap wild animals. Also do some mining, but with the garden and plenty of fish and clams we are able to carry on." Although Joe listed himself as a fox farmer in an Alaska business index as late as 1947, the Ibachs exchanged their Forest Service fur farm permit for a five-acre residence permit in 1938 and concentrated on mining in Glacier Bay.[50]

The Ibachs, however, were among the few fur farmers who eventually gained full title to their land. Between 1935 and 1940, President Franklin Roosevelt, at the behest of the Forest Service, issued at least six Executive Orders removing specified homesite-permit parcels from the Tongass and Chugach Forests. The removal made these homesites available for purchase by the families who lived on them. Most of the several-acre lots surrounded towns where current or former fur farmers now had local jobs and children in school. These included families near the experimental farm at Petersburg, in the Mendenhall Valley near Juneau, at Mud Bay near Ketchikan, and on the outskirts of Wrangell. Although Joe and Muz did not live close to a town, they were well-established in the Gustavus region and in 1949 also received a deed from the government. The couple remained on Lemesurier Island for the rest of their lives. Muz died in 1959. Joe, age 83, took his own life the following year.

Many of the farmers who outlasted the Depression were those who raised mink rather than fox. The fur markets were not saturated with mink pelts, and fashions were shifting. During the 1930s, sleek, short-haired, pale furs became stylish. (Fashion designers and mink advertisers were said to have suggested that long-haired dark furs did not make the wearer appear

Joe and Muz Ibach at home on Lemesurier Island.
Photo by Bruce Black, 1954.

slim.[51]) The new fashions led the Seattle Fur Exchange to update its logo into a map of Alaska framed by a medallion of a fox on the right and a mink on the left.[52]

Earl Ohmer in Petersburg and Gerrit Snider in Wasilla, owners of the two largest mink farms in Alaska, remained solvent through the Depression. Ohmer's Yukon Fur Farm even managed to sell breeding mink to Sweden and Chile in 1936 when the Depression started to lift. On the Seward Peninsula, Percy Blatchford, described by a Biological Survey agent as "the man who pioneered and demonstrated that the white fox could be raised in captivity," kept his farm going by switching from foxes to mink. His daughter remembers the new mink breeders being unloaded from a plane in individual five-gallon cans and settled onto the farm about two miles from Teller.[53] Blatchford's northern neighbor, George Goshaw, stuck with foxes and finally succeeded in breeding a consistent "platinum" strain from blue and white arctic foxes. Fur from these foxes, which would have been described as "dingy" in the previous century, now suited the rising fashion for pale furs. In contrast, government agents on the Pribilofs stuck with their long concentration on breeding the darkest fur possible, even though it worked against them in the marketplace. Pribilof blue fox pelts sold for about $7 less than the average Alaska price during the last half of the 1930s.

Seattle Fur Exchange logo, ca. 1940–1973.
Image courtesy of Dederer Family Collection.

The Great Depression 1930–1940

European War and American Fur Markets

License data from 1941 suggest that only about 40 percent of Alaska's fur farms made it through the Depression. Silver fox farms were hardest hit, with only 30 percent surviving, while 80 percent of mink farms remained in business.[54] But for all fur raisers, 1939 marked the end of a tentative recovery. Just twenty years after the War to End All Wars concluded, Germany invaded Poland and initiated a new and larger conflagration. Fur markets across Europe closed, imports flooded U.S. and Canadian fur auctions, and fur prices again failed. The U.S. government tried to protect American fur farmers by capping imports. Foreign countries were assigned a specific number of pelts they were allowed to send to U.S. markets, ranging from 58,300 for Canada to zero for Germany.[55] But the caps proved unnecessary. Within a year, the near-cessation of trans-Atlantic shipping cut off the flood of European furs. For the next five years, North American auction houses carried almost exclusively U.S. and Canadian furs.

Alaska was 8,000 ship-and-train miles from Berlin but uncomfortably close to Germany's Axis ally, Japan. Less than 1,500 miles separated Attu from Hokkaido. In 1940, the United States began building military installations in Alaska. Jule Loftus found himself frequently called away from Petersburg to places such as Kodiak, where he inspected and certified cattle being raised to feed army troops. He was still the only veterinarian in Alaska and could not keep up with the demand. The governor hired a second territorial veterinarian to do the livestock work. The new hire had recent experience in Alaska and was eager to return. He was Earl Graves, who, after the fur crash of 1930, had gotten a master's degree in animal pathology at the University of Wisconsin. His first wife had died, and he arrived in Alaska in December 1940 with his second wife and two children. He continued to work as a veterinarian in Alaska until his death in 1960, although he never again worked with furbearers.

Meanwhile, Jule Loftus had become increasingly pessimistic about the future of fur farming in Alaska. As a veteran who had been gassed during World War I, he believed that the United States would be inexorably drawn into the new European conflict. Labor shortages would cripple farmers, Pacific shipping would be redirected to the war effort, and Alaska would become increasingly isolated. In the spring of 1941, Loftus resigned his job and moved to Oregon, where he became a part-time cattle farmer, part-time veterinarian. Like Graves, he never again returned to furbearer work.

In Petersburg, Loftus was replaced by James Leekley. After completing a degree in animal husbandry at Oregon State University, Leekley had spent three years working at the national experimental fur farm in Saratoga Springs. Hired by the Cooperative Extension Service solely to run the Alaska Experimental Fur Farm, he had no additional territorial duties. Leekley and his family arrived in Petersburg in July 1941. They would live and work at the experimental fur station for the next three decades, witnessing a series of events that would permanently change Alaska fur farming. The first of these came soon. It was the December 1941 bombing raid on Pearl Harbor.

CHAPTER SEVEN

World War II, 1941–1945

A Nonessential Industry in a War Zone

In 1941, the year the United States entered World War II, the Alaska Game Commission issued 256 fur farm licenses. These licensees, together with fur farmers all over the nation, were on the precipice of the worst half-decade the industry had yet to experience. They would be hobbled by labor shortages and price controls, and they would face an unexpected threat from the development of synthetics for use in cold-weather clothing. The Japanese invasion of the Aleutians would put an abrupt end to fur farming in the island chain. In the territory as a whole, three-quarters of the fur farmers would quit, never to return.

In the years that preceded 1941, Alaskans had come to feel wary of their Pacific neighbor, Imperial Japan. But on December 7, when shortwave radios announced that Pearl Harbor was being attacked, Alaskans were shocked. Within days, the United States had declared war on Japan and Germany, and Alaska had been designated a war zone subject to military regulations. Troops and planes were moved north onto bases that were still under construction. Detachments of soldiers went to work building the Alcan Highway, a 1,500-mile gravel road that would connect Alaska with the States. Steamship companies raised their rates to include war insurance, and ship passengers leaving Alaska were not allowed to board without a military exit permit. Mail to the territory was censored.

Throughout the nation, mandatory registration for the military draft was expanded to include all males, ages 18 to 64, although in 1942, only healthy men, ages 18 to 35, without dependents were actually drafted. Later in the war, older men and those with dependents would also be called up. Almost immediately, Alaska draftees and volunteers began leaving for training camps. Among those donning uniforms were a number of fur farmers and their laborers. The civilians who remained in the territory did not move to fill these fur farm vacancies. High-paying jobs with military contractors were far more attractive than the uncertainty of raising mink or foxes. Women farmers and older men were able to continue, but as the war progressed, Alaskan fur farms began to close at rates higher than during the Depression.

A Farm Lost to Internment

At least one fur farm family suffered an immediate and catastrophic loss as a result of Pearl Harbor. For more than a decade, Ryotaro Urata and his wife Chiyo had owned a mink farm near Wrangell. They had prospered in spite of the Depression because they were among the few who were able to continue selling breed stock. *The Wrangell Sentinel* trumpeted their success on its front pages: February 4, 1938, "Urata Has Successful Mink Year"; December 30, 1938, "Urata Exports Mink Breeders North Japan." The distant farm to which Urata took his breeders in 1938 was located on the Japanese-controlled Korean Peninsula. The land was owned by Urata, and the farm was operated by his brother-in-law. Ryotaro and Chiyo accompanied their fifty breeders on the 43-day trip to Korea, leaving the Wrangell farm in the hands of their oldest son. Before returning, they deposited their profits in a Japanese bank. Although Ryotaro had come to Alaska before World War I and his sons were born in the territory, he looked forward to a comfortable retirement in his home nation and had never acquired U.S. citizenship.

The U.S. government and the Territory of Alaska both encouraged naturalization of immigrant miners, fishermen, and fur farmers who came to Alaska during the first third of the twentieth century. Part of this encouragement consisted of making life inconvenient for those who did not become Americans. For example, aliens were charged extra for hunting licenses and were not allowed to fish commercially. In 1937, Urata had been fined $100 when he was caught fishing for his mink farm. Nevertheless, he had held on to his Japanese citizenship.[1]

But naturalization would not have saved Urata from the fate of all Alaska and West Coast residents who were more than one-quarter Japanese by

Wrangell mink farmers Ryotaro and Chiyo Urata.
Photo courtesy of Masuye Urata.

blood. With the exception of ethnic Japanese women married to white men and ethnic Japanese men serving in the armed services, all ethnic Japanese living on the West Coast, regardless of citizenship, were "relocated" from their homes to internment camps for the duration of the war. By the time the Alaska deportation order came in April 1942, the Uratas' younger son was in the military, but the parents, the older son, and four other Wrangell residents were part of a contingent of more than one hundred people sent from Alaska to a camp at Minidoka, Idaho. All Urata's minks were pelted, and the farm was closed. Although neighbors maintained the Urata family home during the war, the mink farm never reopened. After the war, the sons returned to other occupations in Alaska. And by the time the Japanese bank that held Urata's account sent him a pitifully small postwar restitution, Ryotaro and Chiyo had moved to Seattle.

Aleutian Fox Farms under Double Attack

A larger group of fur farmers for whom 1942 was also disastrous were those in the Aleutians. Aleutian fox farmers had actually begun fearing the loss of their farms well before Pearl Harbor, not because they expected the Japanese to invade but because they felt threatened by the U.S. government. The report by government biologist Olaus Murie on his 1936–1937 wildlife studies in the Aleutians had created a furor in the bureaucracy. After examining more than 1,800 fox scats from forty islands, Murie reported that only 31 percent contained seashore invertebrates, mainly crustacean sand fleas with a small number of sea urchins. More telling was the fact that 57 percent contained bird remains—pigeon guillemots, petrels, puffins, and more than a dozen other species. He watched the adaptable foxes raid upland nests, swim to offshore rookeries, and even "leap across a chasm . . . to the top of a pinnacle where ducks are nesting, then [after eating birds and eggs] clamber down the pinnacle and swim back to shore." One day Murie noticed some foxes perched on the rocks of a cove, watching ducks dive. He reported that when a duck began to rise for air near the rocks, a fox would "jump in and seize it while it is still below the surface."[2]

Murie also counted birds. On islands with foxes, bird colonies were seriously reduced. Cackling Geese, which were reported by Lucien Turner in 1886 to breed by the thousands along the Aleutians from Agattu to Semidi, were so scarce that Murie feared they would "soon disappear from the Aleutian fauna." The Aleutian Canada goose, also once in the thousands, now bred only on Buldir Island, which had never been a fox farm. When Murie circumnavigated the island by boat, he was able to spot only "a few pairs." His report also described Cassin's auklet and the whiskered auklet as "rare."[3]

Conservationists in the Bureau of Biological Survey, already disenchanted by the eagle bounty Alaska had enacted at the request of fur farmers, were deeply troubled to learn of the harm caused by Aleutian fox farms. Under American administrators, foxes had been transplanted onto half of the 175 named Aleutian Islands.[4] Other islands had residual fox populations from Russian days. The agency that had encouraged fur farm development on the islands as an economic substitute for sea otter hunting was alarmed by the devastating effect of imported foxes on island bird life. The Biological Survey now recognized the inherent contradiction in the 1913 executive order that established the Aleutian Island Reservation as "a preserve and breeding ground for native birds, for the propagation of reindeer and fur bearing animals." In 1940, the reservation was renamed the Aleutian Islands National Wildlife Refuge. The refuge was to be administered by the Fish and Wildlife Service, an agency formed by the merger of the Bureau of Biological Survey and the Bureau of Fisheries under the Department of Interior. This new agency was directed to develop regulations based on three principles. First, the "blue fox industry ... will be regulated to avoid destruction of important forms of bird life." Second, "native birds will be studied and attempts will be made to prevent their extermination." Third, "natives ... will be given means to obtain their economic independence."[5] Non-native Aleutian fox farmers, aware of the new government emphasis on native birds and indigenous people, described themselves as being "in a jittery condition."[6]

In keeping with the new regulations, Secretary of the Interior Harold Ickes announced in 1940 that the government would not renew the yearly fur farm leases for sixteen islands, including Amchitka, Agattu, Amukta, Unalga, Kagamil, and Uliaga. To protect birds, all foxes on the sixteen islands were to be removed or eradicated. Alternate islands were proffered to nine displaced village fur co-ops, often at the expense of non-Aleut companies or individual farmers. For example, the Attu Native Community would have to withdraw from Aggatu, but it would be issued a permit for Adak Island—where it would displace the current leasee, the Adak Ranching Company.[7]

The man in charge of the Adak Ranching Company was the same entrepreneur who headed the Kanaga Ranching Company—Harold Bowman.

The Aleutian Island Fur Farmers Association, composed largely of non-Aleut farmers, sent a twelve-page letter of protest to Governor Ernest Gruening with copies to Congressional Delegate Tony Dimond and Secretary Ickes.[8] The letter pointed out that farmers whose leases had been canceled would lose a substantial investment in caretakers' houses, corrals, feed houses, and other improvements. Moreover, the association asserted, these farmers would be able to capture and transfer only a portion of their breed stock. The letter claimed that the replacement islands offered were less desirable in habitat and access. It stated that Murie's assessment was wrong, that foxes foraged only along coastlines, and that the birds they ate were those blown off the cliffs and onto the beach by storms. The association mistrusted Murie, whom the letter described as the government's "so-called expert, [who] intentionally or unintentionally misled the Bureau. To say that [a fox] goes inland or up the mountains or cliffs in search after food . . . is to clearly show that the person making such a claim does not know or understand the habits of the blue fox."

Aleutian farmers also claimed property rights. The government had never before refused to renew an Aleutian lease or permit except for cause. The farmers believed they had an implied right to uninterrupted use of their islands. Most importantly, the association warned, this arbitrary blow to the industry would bring economic disaster to Aleutian Natives and whites alike. The association emphasized the benefits that white farmers with capital had brought to Aleuts in the region. Now these farmers would be afraid to invest, and caretakers and trappers would lose their jobs—all because the Department of the Interior wanted to "conserve some rare wild bird which has no commercial or food value . . . [and] which no one but agents of the Wildlife Bureau or a few wealthy who can afford it, can ever see!"[9]

There is no evidence in the files that Governor Gruening replied to this letter beyond his office's acknowledgment of its receipt. However, he asked for and received written responses to it from the Office of Indian Affairs, the Alaska Game Commission, and the Fish and Wildlife Service—all of which supported the decisions made by Secretary Ickes. All seemed to share the opinion that the powerful whites had exploited Aleuts more than they had benefited them. The Fish and Wildlife service stated that their current policy was to place endangered birds and sea otters first and sustainable Aleut fur farming second. There was no mention of non-Native fox farmers.[10]

In the end, no farmers were actually displaced as a result of this new policy. The whole controversy became moot when the Japanese bombed Unalaska and seized Kiska and Attu Islands in June 1942. All Aleutian residents west of Unimak Island were evacuated on the army transport ship *Delarof* with just a few hours' notice. Houses, pets, and heirlooms were left behind. Blue foxes on islands were abandoned. White families in the Aleutians were able to choose their destinations. But the 880 Aleut evacuees were sent to camps in southeast Alaska—abandoned mines, canneries, and herring salteries—to wait out the war.

Pribilof Fox Farming Suspended

About half the Aleuts transported on short notice were from the Pribilof Islands, which lay 250 miles north of the Aleutians. Shunted to an abandoned cannery and mine at Funter Bay on Admiralty Island west of Juneau, the Aleuts of St. George and St. Paul faced the worst conditions of all the evacuees—empty, poorly insulated barracks with space for families walled off only by blankets; outhouses suspended over the low-tide line, polluting the bay; a mess house with a single stove; no fishing or hunting gear; no church; no clinic; and no school. High school students were sent away to a boarding school in Wrangell. A public health nurse made monthly visits, but measles and tuberculosis soon joined malnutrition, and a cemetery was established. The men were paid $20 per month for splitting wood and digging drainage ditches, but before long, word of good jobs in Juneau reached the camp, and men began slipping away to the nearby town. Government agents—accustomed to thinking of the Pribilovians as wards of the government—tried to keep them in Funter Bay, but soon men were sending back both money and a description of how people lived in Juneau.

In the spring of 1943, Pribilof managers strongly urged the men to leave their jobs and their families at Funter Bay to return to the Pribilofs for a month to harvest fur seal pelts and to provide winter food stores for the foxes. The managers were persuasive, using promises of money, threats of refusing war-end transport home, and appeals to Aleuts' patriotism, stating the harvested furs would clothe soldiers. In the end, they were able to gather a crew to go north. No foxes were trapped, but the harvested fur seal pelts were sold by the Fouke Fur Company and the funds deposited in the government treasury.[11]

By the following summer of 1944, the Japanese had left the Aleutians, and the residents of St. Paul and St. George were brought home.[12] Thirty-five people, mostly elders and children, had died during the two-year internment. When the surviving Pribilof Aleuts returned to their villages, they found their houses had been ransacked by U.S. Army troops stationed on the islands. It was a searing evacuee experience, but one that taught the Aleuts that the United States had labor laws and that U.S. citizens had rights. Although Pribilof men harvested 627 foxes their first winter back home for the usual $5 per pelt, their wartime experience had given them a new point of view. Over the next three decades, they would succeed in asserting their rights to citizenship, fair wages, and control of their land.[13]

Furs in the War Effort

At Funter Bay, government agents had been disingenuous when they told Aleuts in 1943 that Pribilof seal pelts would go to clothe soldiers. This is not because the role of furs in the war effort failed to receive close attention. In 1942, Frank Ashbrook, head of the Fur Division of the newly formed Fish and Wildlife Service, temporarily left his job to act as an advisor to the War Department on the use of furs in war effort. Fur farm journals energetically promoted fur garments for soldiers. An article titled "Washington Committee Designs Fur Lined Aviator's Uniform," showed photos of a pair of cotton overalls, a parka, and "tall Russian boots"—all lined with muskrat fur. The outfit had been designed to omit zippers or other cold-conducting metal. The accompanying mitts had fur both inside and out. Another article advised fur farmers to consider the tactics of the enemy. "Look into last winter and hear Mr. Goebels [sic] begging the German populace for fur to save the lives . . . of her soldiers at the fronts of Russia. . . . Fur, our fur, may be one of the most strategic of raw materials."[14]

At the beginning of the war, Germany had pushed hard to increase its production of farmed furs. After invading Poland, for example, the Germans converted the Warsaw Zoo into a fox farm. But the United States government took another approach: It sought to develop sturdy, lightweight, mass-produced synthetics to replace the heavy, fragile furs used in military garments during World War I. By the end of World War II, the army and air force were using fur only for ruffs. Cold-weather military parkas had a windproof cotton outer layer reversible from olive drab to white and a separate quilted liner filled with mohair and alpaca pile. Soldiers' cold-weather footgear was also synthetic. Rubberized white "bunny boots" had wool and felt sandwiched between their inner and outer layers and were fitted with air valves so that pilots could equalize the pressure between the sealed layers with the changing pressure in the cockpit.

When it became apparent to fur farmers and furriers that the military had no significant use for utilitarian furs, some in the industry argued for the value of luxury furs as a builder of morale and damper of inflation.

> [Fur] lifts the spirits of our womanhood to meet the sorrows and griefs of war and . . . to cheer our fighting forces. . . . Contrast, if you will, the letter of the girl who can flutter herself upon the white page to appear before her hero from the background of fur with the letter of the girl, depressed . . . for lack of those flattering adornments, which nature has decreed the desire, if not the necessity of femininity.[15]

> Since fur coats use no materials needed in the war effort and since every woman in the country wants a fur coat, they are an ideal absorbent of excess purchasing power, which is creating the danger of inflation. . . . More fur coats [for sale] would reduce buying pressure on articles using war materials.[16]

Government Controls on Furriers and Fur Sales

Not surprisingly, these appeals from the fur industry failed to convince government agencies involved in the war effort. In 1941, Congress levied a 10 percent excise tax on furs and then raised it to 20 percent in 1944.[17] In the spring of 1942, the Office of Price Administration set a ceiling on sales of raw furs that froze prices for the next two years. The maximum sale price for mink, fox, or other pelts was not allowed to exceed the highest price paid for a comparable pelt in March 1942. The result was a tight and complex pricing system. Auction houses distinguished five types of silver fox pelts: black, quarter-silver, half-silver, three-quarter silver, and full-silver. In contrast to nineteenth century tastes, all-black pelts were now the least popular and the cheapest. The price rose for each added increment of silver. Pelts with silver guard hairs only on the tail and rump brought less than those with silvery fur from tail to ears. Each of these varieties was then graded for size and quality to determine what a comparable pelt sold for in March 1942. Within months, buyers were all bidding the same price—the March 1942 ceiling. Vocal auctions ceased. Buyers willing to pay the ceiling price for a particular lot of furs submitted their names for a drawing to determine the purchaser. Although the 1942 ceiling prices were higher than 1938 prices, the volume of furs sold was down. Alaska statistics for these years are illustrative.

	1938 Volume	1938 Price[18]	1943 Volume	1943 Price
Blue fox	11,400	$26.50	1,800	$34.00
Silver fox	1,100	$29.50	300	$40.00
Mink	39,800	$11.50	33,700	$12.50

While the Office of Price Administration was regulating prices during the war, the War Manpower Commission (WMC) was regulating labor. In 1942, it declared the fur industry nonessential to the war effort. Fur farmers were not directly affected by this classification, but it slowed sales of pelts because the manufacturers who transformed raw furs into coats and accessories lost their most skilled employees—the furriers who matched the pelts and did the sewing. Furriers (along with jewelers, dishwashers, tailors, bartenders, butlers, dance teachers, gardeners, chauffeurs, astrologers, and two dozen other types of workers) were classified as doing non–war-related work. A furrier who did not voluntarily switch to a new job in a war-essential industry could be involuntarily reassigned by the WMC or even drafted into the army. Draft eligibility by that time had expanded to include men ages 18 to 45. Military deferments were still being given to men with dependents, but the WMC pledged to disregard wives and children if they had to draft recalcitrant furriers or other nonessential workers into the army.

James Leekley at the Alaska Experimental Fur Farm

By 1944, the military draft exemption for men with dependents was abolished for everyone. In Alaska, James Leekley, 34-year-old father of three and the only professional employee at the Petersburg experimental fur farm, received

James Leekley holding a mink.
Photo taken from Agroborealis, October 1970, p. 41.

a notice of new eligibility from his draft board in Saratoga, New York, where he had registered while working at the national experimental fur farm.

During the war, the Alaska Experimental Fur Farm was operated jointly by the University of Alaska and the Alaska Game Commission. Leekley had continued his scientific work, including nutritional studies with mink and foxes. During the war, red meat was in short supply. As a result, it was both expensive and rationed. Fur farmers in the States formed co-ops to buy relatively cheap horsemeat. This meat, from animals that had been displaced by tractors or were past their prime, was also sold in butcher shops during the war.[19]

Aerial view of the Alaska Experimental Fur Farm near Petersburg.
Photo courtesy of Janet Leekley Eddy.

Eugene and Marie Weschenfelder of Spuhn Island near Juneau with their "Atka Platinum Blue Foxes," a pale strain similar to that developed by James Leekley at the Experimental Fur Farm.
Agricultural Experiment Stations, 1968-0069-00097, Archives, University of Alaska Fairbanks.

In Alaska, the price of shipping made imported horsemeat practical only for fur farms located close to major transportation routes. Moreover, Alaska game law forbade feeding moose or caribou to domestic furbearers. Fresh salmon was worth far more sold to a cannery than ground up for mink or fox feed. Cannery fish waste was the only economical source of protein, but it seemed to create nutritional problems. Mink developed "watery hide disease" when fed mainly salmon heads and developed anemia on a diet of halibut heads. Fox pup survival was poor on a variety of fish diets. Leekley tested various fish combinations for these animals with and without the addition of eggs and vitamin supplements. Interior farmers had long supplemented their feed with vitamin-rich lowbush cranberries, picked in the fall and stored through the winter. Leekley worked on a similar additive for coastal farmers—dried tomato skins from canneries in the States.[20]

In addition to nutritional studies, Leekley was making progress in long-term breeding experiments not only with marten but with foxes. Like George Goshaw at Shishmaref and Gene Weschenfelder near Juneau, Leekley realized that the increasing popularity of lighter furs applied not only to silver foxes but also to blue foxes. By crossing and recrossing blue arctic foxes with white arctic foxes, he was able to produce litters in which a third of the pups were a very light blue and worth far more than their ancestors. The Alaska Experimental Fur Farm's "platinum blue" pelts made their debut at the New York fur market in 1943. As did the feed studies, this research had potential economic importance to the territory.[21]

Both the University of Alaska and the Alaska Game Commission were concerned that these scientific experiments would have to be scrapped if Leekley left for the war. They sent a joint wire to Leekley's draft board in Saratoga Springs requesting a deferral. Then Earl Ohmer, chairman of the Alaska Game Commission, sent a separate wire with the same request, signing this telegram with his title as chairman of the Petersburg draft board. Leekley received a deferment and continued his work throughout the war.[22]

The War's Legacy: Sixty Surviving Alaska Fur Farms

No matter how useful the information that flowed from the Petersburg fur experiment station was, it was not enough to keep most Alaskan fur farmers in business. Squeezed by labor shortages, fur taxes, price controls, loss of the international market, expensive feed, wartime shipping rates, and the rise of synthetic cold-weather clothing, only five dozen Alaskan fur farmers were able to hang on through World War II. Some were silver fox farmers, but most were mink farmers such as Gerrit Snider of Wasilla and Earl Ohmer of Petersburg. Only a few—such as Josephine Sather, who would farm Nuka Island until the early 1950s, and George Goshaw, who would sell pelts until 1949—were still raising blue foxes.

Alaskans who kept their fur farms going through World War II were resilient optimists. They reminded themselves that the end of World War I had created an unprecedented upswing in fur sales, especially in the luxury fur trade. They felt sure that Germany and Japan would be defeated, and they looked forward to postwar prosperity. They were right about the economic rebound, but they misjudged the role fur farming would play in the post-war economy.

CHAPTER EIGHT

Post-War Hopes and Decades of Decline, 1946–2000

Cold War, Oil Boom, and the Demise of Alaska Fur Farming

When World War II ended in August 1945, Alaska's few remaining fur farmers looked forward to the recovery of their industry. Price and labor restrictions that had hampered their business were lifted within months. International shipping resumed, and overseas fur markets were expected to reopen. Veterans, some of them experienced fur farmers, returned to the Territory, looking for opportunity and employment. But the Alaska they came back to was not the familiar one they had left.

Most noticeable were all the newcomers. The 1940 census listed approximately 72,500 Alaskans. By 1950, the count would almost double to 128,600. The population mix had also changed. In 1940, Natives made up 45 percent of the population and the military less than 1 percent. By 1950, there would be 25,000 soldiers and 34,000 new civilians in the territory. The proportions would become 14 percent military, 26 percent Native, and 60 percent non-Native civilians. Schools and housing were strained, especially in Anchorage, which almost tripled its population during the 1940s.

Because fur farming and trapping had been Alaska's third largest industry before the war, the Alaska Development Board assumed that many veterans and newcomers would be eager to start mink or fox farms. In 1947, the board published a 40-page booklet, *Fur Farming Opportunities in Alaska*, with general information for beginners and chapters on specific animals and regions. Under the subtitle "Future Is Promising," the booklet noted that "wartime obstacles forced 35% of mink ranches and 25% of the silver fox breeders to quit business during the winter of 1943–1944. . . . [But a] normal demand for fur after the war will necessitate the reopening of those ranches. . . . If there is an extended postwar period of prosperity, there will be opportunities for an even greater number of people in this business."[1]

The Aleutians: Fox Farming Ends, Fox Eradication Begins

In its sections on specific regions, the Alaska Development Board made no mention of one area in which commercial fur farms would not be reopening—The Aleutian Islands Wildlife Refuge. The last permit to transplant and farm foxes on the islands beyond Unimak Pass was granted in 1945. The days of Aleutian commercial blue fox farming were over. Even without government cancellation of fur farm permits, the local industry probably would not have been able to compete with fur farms elsewhere that were more efficient and closer to markets. Despite two centuries of experimentation and despite a few successful decades, farming methods in much of the Aleutians had remained rudimentary.

A view of the last days of Aleutian farming comes from ship captain George DeVenne. In 1939, DeVenne was working in the western Aleutians and described island farms where delivery of lumber was rare and expensive: "There were no feed houses, nor pretence toward the pen-raising of foxes." Cod delivered to islands without feed houses was quickly plundered by ravens, eagles, and seagulls. For the most part, foxes had to find their own food. Few of the islands had good harbors, and trappers working the forty-day prime fur season often ended up stuck on an island for several months, waiting for weather that would allow a ship to pick them up. DeVenne recalled that he and two partners had once tried trapping Adak Island by using a boat to move from harbor to harbor around the large island. But the coastline offered only marginal protection from winter winds. At one point, he reported having to "put down over eleven hundred pounds of anchor gear to keep my little ship from being blown ashore." During the forty-day season, the three men managed to trap for only eighteen days. Instead of the

150 pelts they hoped to secure, they brought home only seventy-five. It was DeVenne's opinion that the commercial future of the Aleutians lay not in furs, but in salmon, cod, mackerel, and king crab.[2]

After the war, none of the members of the Aleutian Islands Fur Farmers Association returned to their farms. Most of them, like Harold Bowman of Kanaga Ranching, simply left Alaska, never to return. A few, like Nicholas Bolshanin of Sitka, pursued other work in the territory. The only member known to have resumed any type of fur farming was Captain C. T. Pedersen, who developed a mink and silver fox farm in Quebec, Canada. Pedersen also engaged in a two-decade-long struggle through the Foreign Claims Settlement Commission for restitution from Japan, whose attack had destroyed his Aleutian Fur Company warehouse, merchandise, and motor vessel on Attu.[3]

In 1949, the Fish and Wildlife Service began working to eradicate foxes from important bird nesting islands, beginning with Amchitka. The task was not easy. Trapping alone was insufficient to clear an island. As one trapper explained, "After the first few days of trapping, those which are not caught become very cunning."[4] Poisoned meat was more effective, but there was a risk that the poison would kill eagles and other carrion feeders. A decade of work to clear an island was not unusual. It was eleven years before Amchitka was declared free of foxes. Eradication begun on Agattu in 1964 was not complete until 1979. However, by 2008, more than forty Aleutian Islands were emptied of foxes. Rebound of the bird populations on these islands was gratifying. The Aleutian goose, which had required a special breeding program to help it survive, was removed from the endangered species list in 1990. Today, 40,000 Aleutian geese nest on eight or more islands. Currently, about thirty Alaskan islands, most of them in the Aleutians, still harbor non-native foxes—less than 7 percent of the estimated 455 islands throughout Alaska that at one time or another functioned as fox farms.[5]

The post-war years were good for Aleutian birds but difficult for Aleuts. They returned from relocation camps in southeast Alaska to find their villages in shambles. Attu residents, returning from prison camp in Japan, were actually barred from their island. The U.S. government was unwilling to help rebuild, provide services, or even transport Attuans back to such a distant part of the Aleutian chain. Only Atka, Akutan, Nikolski, and Unalaska were resettled. Aleuts who wanted to trap residual fox populations on nearby islands were often blocked by red tape because the military had withdrawn a number of islands from the refuge during wartime and was slow to relinquish them when peace returned. Only a few post-war Aleutian blue fox pelts were sent to market. The days when three men could trap an island farm for six weeks and return with over 300 pelts became the subject of stories told by the elders.[6]

Pribilof Fox Farming after WWII

North of the Aleutians in the Pribilof Islands, post-war fox farming was also in trouble. Frank Ashbrook, head of the U.S. Fish and Wildlife's Fur Division, was deeply concerned about the poor prices the government was getting for their blue fox pelts. In 1946, the harvest was good, and the official bulletin, "Furs Shipped from Alaska," reported that Pribilof blue fox pelts sold for the same price as those from private farms—$35 each. However, Ashbrook's printed correction on the report in his file indicated that the government actually received only $9.50 per pelt.[7] By the following year, the wartime backlog of buyers' orders had been filled, and not only did the price for blue fox pelts fall precipitously but auction houses had trouble selling all the pelts they received. Private farmers in 1947 received about $11 per pelt. The government received $8.50. In 1948, only a few of the Pribilofs' approximately 1,100 pelts sold; most were held over at the auction house for the next several years. The 1948 Pribilof price was a dismal $4.22—less than the $5 per pelt paid in wages to the Aleuts for the previous three decades.[8] The government cut pay for local workers to $3 per pelt.

The problems, Ashbrook reasoned, were fashion and competition. Furriers were buying light-colored, short-haired furs to suit new postwar styles. Pribilof foxes, bred for fashions of the 1920s, had fur that was "too open, too coarse, too dark, and lacks luster." Ashbrook was not just comparing Pribilof fox fur to that produced by private farmers in Alaska or in the States. He was measuring it against the "silky, dense, light silvery blue fur" produced in Finland, Sweden, and especially in Norway.[9] He was apparently unaware that 450 miles north of the Pribilofs, George Goshaw at Shishmaref was already producing light blue foxes by crossing selected offspring from years of white-blue matings.[10] To improve Pribilof pelts, Ashbrook proposed restocking St. Paul Island by trapping all the native foxes and replacing them with pale, high-quality blue foxes from the Petersburg experimental fur farm or even from farms in Norway. He estimated the cost of his proposal at $42,000.[11]

Although discussed for several years, Ashbrook's plan never came to fruition. Blue fox prices stayed low, and by 1947, Pribilof Aleuts, working with legal counsel to obtain independence and self-government, were finding $3 per pelt meager pay. After 1950, trapping decreased rapidly with most government sales coming from the backlog of fox pelts held over from earlier years. The commercial harvest of Pribilof foxes ended in 1955, when twenty-five animals were trapped and pelted. The pelts sent to auction earned the government a total of $50 after wages.[12]

The end of fox farming in the Pribilofs meant that the fox population was unchecked and the feed houses were empty. On St. George, the villagers found themselves inundated by hungry foxes prowling the village to pillage chicken yards, garbage cans, and unattended leather gloves or boots. The villagers began their own fox eradication program in 1953. Using leg-hold traps, they caught and destroyed 342 animals the first year without discernible effect.[13] After several more years of vigorous trapping and shooting, the fox population finally shrank to the approximate number that had been present before the Russians arrived, and the foxes were again able to support themselves on natural foods from the island's ecosystem.

Southeast Alaska: Native Rights and New Forest Service Policies

Far to the east of the Aleutians, another federal agency was also dealing with issues that involved Natives and fur farming. In 1946, investigators for the Commissioner of Indian Affairs held hearings in southeast Alaska and published a report titled "Possessory Rights of the Natives of Southeastern Alaska." It contained testimony from eighty-eight Tlingit and Haida witnesses who described being excluded from traditional subsistence islands by white men holding various Forest Service permits. At least twenty-three fox farms were mentioned in the testimony.[14]

As early as 1923, a Forest Service inspector had expressed concern about fur farms and "the inevitable conflict with . . . Indian rights."[15] Fur farmers holding permits were prohibited from disturbing Indian cemeteries, totem poles, or gardens. But once the "no trespassing" signs went up, entire islands—regardless of traditional gardens, berry patches, fishing spots, or smokehouses—were declared off limits by the permitees.[16] The sense of white entitlement was strong. In 1923, when a Ketchikan Forest Supervisor required the applicant for Harbor and Round Islands to obtain permission from local Indians, the applicant's lawyer forwarded quitclaim deeds signed by Sumdum Charley and Sumdum Ben. Accompanying the documents was a sarcastic letter stating that although his fur farmer client believed these Indian claims were "based upon mythological ancestries," the client, "fully expecting these islands to be haunted by the ghosts of very long dead Indians . . . was willing to take a lease subject to all such bunk."[17] By 1946, such sarcasm was considered offensive, and the testimony of southeast Alaska Indians helped to convince Congress to pass the Indian Claims Commission Act, which allowed all Alaska Native groups to sue the government over claims to land and other issues.

However, as early as 1940, the Forest Service had—for reasons that had nothing to do with Indian land claims—already become reluctant to issue island fur farm permits.[18] Citing a history of financial failures marked by empty promissory notes and abrupt closures, foresters developed a policy of encouraging fur farmers to leave the islands. In a letter to a district ranger, the head of the Alaska Forest Service, B. Frank Heintzelman—who would eventually succeed Gruening as governor—explained why he had rejected a new permit for Kanak Island. The island was remote, had a poor harbor, and was too large for a farmer to keep track of foxes. It was no surprise that three previous farmers had failed to make a living on the island. But there were also overriding policy issues for rejecting the application. These were based on social concerns:

> It is definitely not in the public interest to encourage occupancy of these isolated islands by families with children unless the enterprise can stand the expense of boarding the children in the nearest town for nine months of the school year. . . . Neither can the Federal Government usually provide mail facilities. Infrequent human contacts are likely to be detrimental to the permittee and his family. Lack of such community facilities, services and contacts has led to appalling backwardness of the people in certain isolated sections of the United States [which] all who have considered Alaska's future are agreed [we] should seek to avoid.[19]

Charles Rudy chose to farm near the town of Juneau. At Rudy's Ranch in the Mendenhall Valley, he raised fox, mink, and marten from the 1920s through World War II.
Photo courtesy of Jim Geraghty.

Heintzelman also noted that the territorial veterinarian recommended against allowing foxes to range free. Only with pens could feeding, breeding, and parasite problems be controlled. In its section on southeast Alaska, *Fur Farming Opportunities in Alaska* would describe only pen farming and would emphasize the new money crop—light-colored mink.

To encourage longtime fox farmers to leave their islands, the Forest Service laid out five-acre homesites along the edges of coastal towns for island families willing to move. The homesites offered adequate room for pens, and settling near a town made economic sense in part because farmers could earn supplemental income. And there were other advantages—regular marine routes for delivery of feed, public cold storage, schools that bused students to town, and perhaps the benefits of community life as espoused by Heintzelman. Forest Service fur farm permits dropped from about one hundred in 1940, to fifty in 1945, and to eight in 1955.[20] When farmers abandoned islands in southeast Alaska and Prince William Sound, the blue foxes that were left behind soon died from inbreeding, starvation, disease, or predation by bears.[21]

By the 1950s, scarcely any southeast fox farms remained, not just because of pressure from the Forest Service but also for economic reasons. Fur prices were shutting down fox farms all over the territory. In Alaska, the average price for fox pelts in 1947—blues $11, silvers $30—had by 1951 fallen substantially to $8 and $10, respectively. Imports of silver fox pelts from Canada to U.S. markets were almost nil, and domestic silver fox pelts were auctioned for a quarter of the $40 that it cost to raise each animal. In contrast, mink was on the rise, moving in the same four-year period from $23 to $28. Of the thirty-one fur farmers active in Alaska in 1951, most were raising mink. The Alaska Development Board had correctly foreseen that mink would overshadow fox, although it had misjudged the overall recovery of the fur market.

Post-War Culture and Fashion Changes

In the 1950s, not only did demand for luxury fox fur deteriorate, but sales of utilitarian furs also decreased. The chemical industry had steadily improved the synthetics that created lightweight, sturdy, cold-weather clothing. Even Eskimos chose to work in quilted coats with lofty, resilient fillings and save their hand-made fur parkas for ceremonial occasions. The utilitarian market had traditionally dealt in cheap furs. Prices for muskrat, skunk, weasel, and other low-end furs dropped as the popularity of synthetics rose. Moreover, in the early 1950s, the market for inexpensive furs suffered a second blow.

Fashion had long tempted unscrupulous furriers and sellers to transform cheap pelts into forgeries of expensive skins. Their practices were mocked as early as 1923 in a collection of satiric cartoons titled *Illustrated Catalogue of Fur Bearing Animals*. In it, a professor discusses a wild Iceland fox while his pointer rests on a sheep, a hunter stalks a black genet (a tomcat on his neighbor's fence), and rabbits run for their lives from a "seal" hunter.[22] Skilled teams of furriers could shear, pluck, dye, bleach, tip, and glaze the hair of almost any pelt. They created stripes or leopard spots using a feather or a spray gun. They "electrified" furs using an iron and a brush to apply a solution that made old fur appear new. The sellers then created a captivating name for the result and a price tag to match. A substantial fringe of the fur industry was strictly "buyer beware."

But consumers objected when their fur coats shed, changed colors, or caused dye-induced rashes. Congress heard the complaints, and in 1952 passed the Fur Products Labeling Act. The law required sellers to show on all labels, advertising, and invoices the actual name of the animal producing the fur. In addition they had to inform the buyer if the fur had been dyed or bleached. Government publications printed lists of previous trade names with new approved descriptions. Rabbit, which had been marketed under more than thirty names, including mink, electric beaver, Hudson seal, and Baltic fox, now had to be labeled mink-dyed rabbit, beaver-dyed rabbit, seal-dyed rabbit, or fox-dyed rabbit. Goat hides marketed as wolf had to be renamed wolf-dyed goat. Sellers of Alaska sable had to admit that their fur was from a skunk with its white strip obliterated by dye.[23] Auction sales of rabbit, goat, skunk, and other cheap furs waned after 1952. Garments created from disguised furs were harder to sell once the manufacturing details were revealed.

In contrast to the market for cheap furs and long-haired foxes, the market for short-haired luxury furs transiently improved in the decade after the war. One of these furs, popular for coats, was chinchilla. This soft gray fur came not from "chinchilla rabbits" but from the small South American rodent for which the rabbits were named. For many years, Andean countries would not allow chinchillas to be exported because, at the beginning of the twentieth century, the European fur trade had decimated their numbers. However, in 1923, an American mining engineer working in Chile collected eleven animals and managed to convince the government to allow him to take them back to California. By the time WWII was over, generations of offspring had transformed the eleven into an ample supply of American-bred chinchillas for sale to fur farmers. In Alaska, at least two post-war chinchilla farms were started in the Anchorage area.[24] Because a pair of chinchilla breeders cost $1,200, these farms were, of necessity, small operations. A chinchilla cage does not require much room, so in addition to those reared on farms, some chinchillas were simply raised inside urban houses.

But in post-war Alaska, the big-money luxury fur was not chinchilla but "mutation" mink. These mink pelts in unusual colors were not dyed. Instead they came from animals whose ancestors were mutants born with fur that ranged from white to cream to tan to blue-gray or red-brown. Prescient fur farmers before the war had begun altering their breeding programs not only to produce light-colored fox but also to produce pale mink by seeking and nurturing mutations, which in earlier eras would have been discarded. Careful breeding after the war resulted in new color strains—blue-iris, dawn, lavender, mahogany, palomino, pastel, pearl, platinum, sapphire, and a half-dozen others sold under glamorous trademark names, such as Lutetia (gunmetal gray) or Arcturus (lavender beige).[25]

The first specialty mink coat (albino-white) was assembled in 1943. But sales of the new colors did not take off until after the war. In the late 1940s, demand increased quickly, but fur in the new colors was still in short supply. For a few years, mink farmers who had developed mutation herds sold more breeders (at $300 to $4,000 per pair) than pelts. Pelt prices rose in response to this relative shortage. In the late 1940s, a full-length platinum mink coat, which required 60 to 80 pelts, sold for $18,000.[26] The average price for Alaskan pelts (which included both natural and specialty colors) reached $45 in 1948. Unfortunately for Alaskan mink farmers, and for mink farmers all over the nation, 1948 marked the peak price for mink pelts in the twentieth century.[27]

Why did the luxury fur market eventually slump and why did so many mink farms fail in the 1950s? It was not that elegant women had given up wearing fur. Magazines pictured movie stars like Grace Kelly, Sophia Loren, and Elizabeth Taylor in pale mink stoles and coats. And it was not because mink farmers were unable to organize. A powerful cooperative, the EMBA (Mutation Mink Breeders Association), staged national advertising campaigns, worked hard to control the fur market, and eventually purchased the Seattle Fur Exchange. The problem, according to Victor Fuchs, author of *The Economics of the Fur Industry*, was related to social change and economic choice.[28]

The post-war period was generally one of prosperity, especially for families who had suffered during the Depression. The percentage of Americans in the middle and upper-middle classes rose. These rising blue- and white-collar families bought cars, suburban homes, dishwashers, automatic washing machines, and clothes driers. For conspicuous consumption, they favored fancy cars and vacation cabins over the furs and diamonds purchased by the smaller upper class. In the United States, there was also a regional shift of population from New England and the Midwest to the Pacific Coast—most notably to California where weather was warm and clothing styles were casual. Moreover, the fur industry failed to adopt marketing ideas that were becoming popular with Americans—installment plans, manufacturer guarantees, an advertised price that included tax (20 percent, in the case of furs), and arrangements for servicing, such as cleaning fur garments or providing summer storage.

Many farmers in leading fur farm states—Wisconsin, Utah, Oregon, Minnesota, and Idaho—were in difficulty after the war. They blamed not demand but a glut of imported furs, especially from Russia. Senator Joseph McCarthy, from the fur farming state of Wisconsin and a celebrated hater of Communists, successfully sponsored a bill to ban the "importation of ermine, fox, kolinsky, marten, mink, muskrat, and weasel furs" from the U.S.S.R. or Communist China. However, the ban did not appear to have a significant effect on fur auction prices.[29]

In post-war Alaska, fur farmers were in deeper trouble than even their competitors in the States. Shipping rates for incoming supplies were high, as were airfreight charges for furs headed to the Seattle auction.[30] Moreover, labor was hard to find. Alaska had become a vital defense link in the Cold War. Fur farmers who needed help with pelting or for vacation relief could not compete with the cost-plus contracts and overtime wages paid during the construction of military bases, radar stations, communication sites, and housing for Alaska's growing population. The fur farm boom of the 1920s had faded into the past, and the construction boom of the 1950s had moved to the forefront.

Fur Farms under Statehood

By the time Alaska became a state in 1959, very few fur farms were left. In 1951, twenty-four fur farmers held permits in the Tongass and Chugach National Forests. By 1959, only two were left. And the Alaska Game Commission, which expired with statehood, issued only seven licenses in its final year.[31]

Two of the largest mink farms in the territory ceased production just before statehood. One of these, near Ketchikan, had been operated by Ernest Anderes for thirty years. At one time the farm had as many as 3,000 mink in pens. Anderes, who wrote the section on "Fur Farming in the Ketchikan District" in *Fur Farming Opportunities in Alaska*, pelted all his animals and closed in 1957. An even larger mink farm, the Yukon Fur Farm near Petersburg, also went out of business in the mid-1950s. Earl Ohmer, the company's

Mink pen buildings at the Ernest Anderes farm near Ketchikan.
Agricultural Experiment Stations Collection, 1968-0004-02890, Archives, University of Alaska Fairbanks.

founder, died in 1955, having already sold his share in the farm to a partner, Jess Ames. At about the time of Ohmer's death, Ames pelted all the animals and put the property on the market. The Yukon Fur Farm was probably capable of handling 4,000 mink and several hundred foxes. It included seven 200-foot-long buildings for mink cages, three large buildings for foxes, a house, a warehouse, and numerous outbuildings. For a few years after it was offered for sale, the buildings remained empty. Then just after statehood, Harold and Ethel Bergmann purchased the farm and restarted it on a smaller scale. The Bergmanns were unable to build the business back to its former size, and after a few years gave up the enterprise.[32]

Anderes' and Ohmer's farms were the last large fur farms of Alaska. After statehood, only small partnerships and family operations survived, a trend that had been predicted decades earlier by the director of Alaska's Extension Service: "Fur farms . . . small enough so that families may operate them without outside help may [provide] a livelihood for many. . . . Larger fur farms . . . have not been so successful."[33] In the new state of Alaska, whenever fur prices rose, there would be a small swell in the number of fur farms; when prices fell, there would be a trough.

As the fur industry sputtered along during the decades between 1960 and 1980, federal and state agencies intermittently tried to lend a hand. In the mid-1960s, the Bureau of Indian Affairs sponsored a mink farm as a new enterprise on Nunivak Island, thirty miles offshore from the Yukon-Kuskokwim Delta. The island is large (1,630 square miles), and the islanders had a well-established reindeer herd. They planned to use scraps from reindeer butchering in addition to fish in preparing mink food. Two villagers were sent to Petersburg to train at the Alaska Experimental Fur Farm. They returned with 25 black mink, but after they got home, they found that maintaining the pens and providing year-round good nutrition was a struggle. Worse yet, mink pelt prices were falling and the only recompense the men were to receive for their work was the profit on fur sales. After a year or two, they gave up and abandoned the effort.[34] By 1966, mink prices languished below $20, and there were only three mink farms and one fox farm left in the state. The Alaska Division of Agriculture was concerned about this collapse in fur farming and commissioned a study of the potential for farming mink in Alaska. Their consultants, the Fur Breeders Agricultural Cooperative of Midvale Utah, concluded that only large mink ranches, with an initial stock of at least 1,000 females and 200 males, would be feasible. Moreover, such a farm would need to be supported by a fleet of fishing boats to supply much of the estimated 400,000 pounds of feed a farm of that size would require.[35] Not surprisingly, the report spurred no new development.

Alaska's Experimental Fur Farm in the Jet Age

Despite the near-disappearance of private fur farms, the experimental fur farm at Petersburg managed to keep its doors open for the first thirteen years of statehood. After the war, James Leekley continued to perform diet and breeding experiments with arctic foxes, marten, and mutation mink. Then in 1970, he undertook a new and elaborate task for the U.S. Air Force: a test of the effect of sonic booms on farmed mink.

Fur farming, rooted in the eighteenth century, and the aircraft technology of the twentieth century had found themselves in conflict and confronting each other in courtrooms. In Alaska, complaints about aircraft noise causing animal losses had been voiced by fur farmers as early as World War II. Ed Opheim, who farmed an island near Kodiak, alleged that in 1942, fighter planes swooping low over his blue fox pens caused the mothers to eat their pups, putting him out of business. Later in the decade, three mink farmers near Anchorage did more than complain. They sought $32,000 in compensation for 452 mink they claimed were lost in 1941 when females, frightened by low-flying military aircraft, destroyed their litters. The farmers stated they were forced to quit raising mink and had to sell their farms at a loss. A bill to compensate the three farmers passed Congress in 1947 but was vetoed by President Truman. A few months later, Congress passed a second bill that became law. It gave the farmers permission to sue for damages in Alaska District Court.

Records do not show that any of the Anchorage mink farmers actually filed a case or received payment from the government, even though they had an encouraging legal precedent in their quest for compensation.[36] In 1937, Alaskan D. E. "Duke" Stubbs had successfully sued the federal government for noise damage on his fur farm. Stubbs raised foxes on private land surrounded by Mt. McKinley National Park. He was a well-known fur farmer in the Interior and on the Kenai Peninsula, where he wrote colorful advice to readers of the "Friday Fur Farming" section of the *Seward Daily Gateway*.

Post-War Hopes and Decades of Decline, 1946–2000

Duke E. Stubbs (left) in front of his cabin in McKinley Park with chicken wire pen in foreground. Mrs. Stubbs on right, Kitty Graves, the veterinarian's wife in center, ca. 1929.
Photo courtesy of Mary Graves Zahn.

(Sample quote: "By scaring the fur farmers of the world nearly into a fit, the makers of so-called worm cures gypped them out of millions."[37]) Stubbs claimed that park employees with horses and dogs repeatedly traipsed across his land without permission and that the noisy commotion caused his foxes to fail to breed. The federal court awarded him $50,000 in damages. He took the money to the States, where he died the following year without re-establishing his fox farm.[38]

During the 1960s, two decades after the Stubbs case, at least twenty-five mink farmers in various states successfully sued the federal government for animal damage caused by aircraft noise. But these farmers were not concerned about low-flying airplanes; they claimed the harm came from sonic booms. They asserted that the explosive bang and the sharp increase in air

James Leekley in front of sonic boom simulator at the Alaska Experimental Fur Farm, 1970.
Photo taken from Agroborealis, February 1970, p. 11.

pressure that occurred when a plane "broke the sound barrier" panicked their female mink into destroying their litters. The Air Force had doubts about whether the sonic booms were actually causing mink losses. At the beginning of 1970, eleven additional mink farmers filed sonic-boom suits in federal court.[39] The Air Force needed scientific proof about what happened inside mink dens when a boom sounded.

Petersburg was chosen as the site for the $100,000 experiment, which involved 650 pregnant mink, a pair of Air Force F-4 Phantom jet planes, and a sonic boom simulator (a giant two-ton amplifier). The mink, housed in cages wired for sound, had movie cameras trained on them. About half the test animals were settled in a quiet location eighteen miles from the farm. Soon after the females delivered their litters, the jets were called in to make three high-speed passes over the cages to generate sonic booms. At a separate location, other females were subjected to three milder booms from a simulator when their kits were born. The rest of the pregnant and birthing mink stayed at the experimental farm to act as controls. The outcome satisfied the Air Force: No harm came to the litters. Mink mothers did not panic and resumed their normal activities within two minutes of a boom. The kits from all three groups were evaluated at fifty days of age and found to have similar weights and apparent health. Moreover, a similar number from each group survived to adulthood.[40]

The supersonic jet test turned out to be the last well-funded experiment at the Petersburg farm. Alaska's military expansion was ending, and its oil revenues were just beginning. Fur farming was now far down on the list of productive industries, and state experts did not expect it to revive. The University of Alaska closed the Alaska Experimental Fur Farm in 1972. James Leekley, who had headed the experiment station for more than three decades, retired, although he continued to be an expert source for reporters writing about Alaska's fur farms until his death in 1988.

The Final Decades: Mutation Mink, Silver Fox, and PETA

Alaska was not the only state with failing fur farms. American fur prices were so low in the late 1960s and early 1970s that the term *mink depression* appeared in trade publications, and about half the mink farms in the United

States went out of business. As noted, fox farms were already at low ebb. Demand for blue fox fur had disappeared, and although silver fox prices had wavered upward, they never fully recovered after the war. Part of the problem for fox farmers was that the new svelte styles favored short-haired pelts. Part of it was that customers had discovered that fox fur was more fragile than mink. Fox coats were prone to developing worn patches, especially if a woman repeatedly wore her coat while driving a car, and more women than ever were driving.

Sales of mink coats, although shorter-haired and sturdier than fox, also suffered in the 1960s and 1970s. The fact was that many women were simply not wearing furs. Although Liza Minelli, Raquel Welsh, and Diana Ross appeared in mink ads, the television and movie stars of their era were more likely to be photographed in bikinis than in fur coats. Brigitte Bardot was certainly not photographed in mink. At age 40, she was just about to retire from film and become an animal rights activist.

People for the Ethical Treatment of Animals (PETA) did not organize until the 1980s, but opposition to wearing fur on moral grounds probably affected mink demand by the early 1970s. The notion that it was cruel to use animals for fashionable clothing was not new. The files of Frank Ashbrook contain clippings from New York and Washington, DC, newspapers on the subject. Several were dated 1923, including "Fad of Wearing Fur All Year Threatens Animals' Extinction" and "What a Deformed Thief This Fashion Is." One titled "Fur-Bearing Woman Taking Heavy Toll from Our Wildlife" in particular decried the frippery of fur trim on boots, sunshades, and night gowns.[41]

The concern of these few people gradually became more widespread. Beginning in the mid-1920s and continuing for more than three decades, the American Society for the Prevention of Cruelty to Animals sponsored an annual contest with a cash prizes for inventors of humane furbearer traps.[42] About the same time, news articles and government reports on fur farms began to emphasize that in order to produce rich fur, the animals were well-fed and carefully tended. Writers described the modern use of painless methods for killing—electrocution, carbon tetrachloride nasal drops, or carbon monoxide.[43] But the greatest change in attitudes came after World War II. In the 1920s, many people who had protested the use of luxury furs still accepted the necessity for utilitarian fur garments in cold climates. With the availability of warm synthetics in the last half of the twentieth century, opponents of wearing fur felt that killing fur animals for cold-weather clothing was neither necessary nor acceptable. Protests in the 1970s and 1980s gradually became more confrontational and doubtless influenced buyers.

Despite these challenges, mink sales managed a partial rebound from the doldrums of the late 1960s. One business that bet on the market's improvement was the Seattle Fur Exchange. In 1973, it expanded into a new 40,000 square-foot building and installed an early IBM computer. In the fall of that year, the business was sold for $750,000 to the EMBA mink breeders association, which had been supplying most of its pelts. The logo with the silver fox and mink medallions disappeared, more fur farm associations bought in, and market rivals like Hudson's Bay Company retired from selling fur. The Seattle auction house, under the name American Legend Auction, became the largest in the Western Hemisphere.[44]

In Alaska during the late 1970s, when mink pelt prices rose from less than $20 to more than $40, state and Native organizations again showed an interest in fur farming. Two new studies and a how-to booklet appeared: "Feasibility of Fox Farming in the NANA Region," "Fur Farm Production and Potential in Alaska," and "A Village Fur Farm."[45] None of this information led to a surge in raising furbearers, but a new state program did help start at least four farms.

In 1978, the legislature, with oil revenues at its disposal, appropriated $40 million to establish the Alaska Renewable Resources Corporation (ARRC). The corporation provided loans to develop Alaska businesses that were based on principles of sustainability. The first fur farm applicant, Jim Rice, requested $100,000 to raise mink and foxes. Rice, who convinced the corporation that the pelts and breeders generated by farming were a viable renewable resource, received his loan in 1979. However, within a few years, his farm failed and he ended up in bankruptcy court entwined with another failed fur farmer. Rice's default was one of several publicized by critics as a reason to dissolve the ARRC.[46]

The Rice fur farm was almost the last in Alaska and would have represented a sad ending to an industry that began with great hope. Fortunately, at least one fur farm remained in business for another decade. It was a silver fox farm operated by Irene and "Rusty" Christie in North Pole, Alaska, near Fairbanks.

Rice had bought many of his silver fox breeders from the Christies, who had started their C&M Fur Farm in 1968. Fourteen years later, their enterprise was featured as a success story in *Alaska Farm Magazine.* During the 1970s, the Christies worked hard to develop a herd that included both standard silvers and silver foxes in specialty colors (platinum, pearl, and amber). For several years during the 1970s, they were the only fox farmers in the state. Their work finally paid off when silver fox prices bumped upward in the early 1980s—an economic effect that some credited to Nancy Reagan's intransigence in the face of those who criticized her fur coat.[47]

In both 1980 and 1981, a silver fox pelt from the C&M Fur Farm brought the top price ($400) at the Seattle Fur Exchange. Fox farmers outside of Alaska took note of this achievement, and C&M breeding pairs began to sell for as much as $1,800. The cost of raising a fox for a year in those days (exclusive of the owner's labor) was between $75 and $130, so the profit from the Christies' "seven day a week, 365 day a year job" was satisfactory. In 1988, when Rusty died, Irene sold the stock and equipment to the Whitestone Farms near Delta Junction. Whitestone continued to raise foxes until about 1993, when the pelt price once again dropped below the cost of raising a fox in Alaska. Both their fox farm and an era of Alaska fur farming came to a close.[48]

Alaska Fur Farming: Dreams and Realities

For more than 200 years, Alaska had seemed like the ideal location for raising furbearers—cold winters, plenty of natural food, and acres of sparsely inhabited land. Andrean Tolstykh saw the possibility of filling "empty" islands with blue foxes. M. L. Washburn and his partners envisioned a flourishing company of farms. James Judge identified the cause of the Pribilof fox decline and designed a program to improve the government's fox harvest. Nicholas Bolshanin foresaw fox farming as economic salvation for Aleuts. William Forbes saw skunk farming as a personal road to riches. Gerrit Snider and Earl Ohmer believed in the future of large mink farms.

But despite many successes, farming furbearers in Alaska was always problematic. The great boreal forests of Alaska are indeed cold and cause furbearers to develop thick winter coats, but the boreal forests do not support dense animal life because food supplies are limited and cold weather requires high energy consumption for nonhibernators. The northern seas, on the other hand, are bountiful, but transferring seafood to land animals requires hard work and money. Moreover, the coldest areas of the state are at the ends of long and expensive transportation routes. In the southeast Alaska rain forest, where supply routes from Seattle were comparatively short, transplanted furbearers developed thinner coats that were prone to matting, and they sickened from parasites that thrived in the warm damp soil. In the Aleutian Islands, which were rich in natural food on beaches and on the hills used by ground-nesting birds, farmers discovered that foxes could destroy their own food supply in a few years.

As for the land, it proved not to be as "empty" as it seemed. Alaska Natives, who ranged widely for subsistence, asserted rights to land they used during various seasons of the year. Conservationists, speaking on behalf of birds, lobbied for clusters of islands and swaths of mainland to be set aside as refuges. Moreover, Alaska's largest landowner, the federal government, tended to be stingy with its property, and fur farmers were restricted to revocable leases or permits that required active use and yearly payments.

And there were the voices of false confidence: "Anyone willing to do a little work can farm," "No need to study books about vitamins or inbreeding or sanitation," or "Free-ranging island foxes stay healthy." The great Alaskan innovation of fenceless island farming regularly failed in the hands of unprepared farmers. Some were ignorant of the basic tenets for raising animals; others were negligent; and more than a few were preoccupied with alcohol stills or other business. Several hired incompetent or dishonest caretakers. Some island farmers lost their gain to poachers.

But it was the market that sounded the death knell. Alaskans raising furbearers knew that the luxury market could be fickle. Nevertheless, its downturns always seemed to come as an unpleasant surprise. Moreover, fur farmers everywhere were unprepared for the worldwide upheavals that affected the sale of both utilitarian furs and luxury pelts: two world wars, a major depression, and the greatest blow of all—the development of synthetic fabrics that were warmer, lighter, and sturdier than fur. The wearing of furs had extended back to the third chapter of Genesis. It is understandable that farmers failed to envision a time when utilitarian furs would barely be needed or when luxury furs would lose their fashion cachet and become a target for animal advocates. They probably also could not imagine that Alaska's farmed furs would be eclipsed in the U.S. market by furs from Canada and from states such as Wisconsin, Minnesota, and Utah. Nor could the farmers imagine that the international market would be dominated by Scandinavia, Finland, Poland, the Netherlands, Russia, and China. Alaska's scattered "mom-and-pop" farms and even its biggest minkeries were unable to compete with modern, large-scale fur farms in the contiguous United States, where overhead was lower, transportation cheaper, and labor less expensive.

Yet, in spite of the failures and the industry's eventual collapse in Alaska, fur farming was for decades a vibrant enterprise that supported merchants, shipping companies, and fur buyers. As the third-largest industry during a period when the price of gold was low and salmon runs were poor, it helped many Alaskans through hard times. It allowed men and families the satisfaction of being their own bosses, working to tame wild animals, pioneering a new industry, and dwelling in the midst of natural beauty.

As we look back at the fur famers' efforts, we get a glimpse of people struggling to make a living in the north by what appeared to be a viable proposition, supplying a local, renewable resource to the world market. Their efforts were part of a historic and continuing quest for sustainable enterprises suitable to the climate—occupations that allows Alaskans to flourish within the great landscape.

ENDNOTES

Abbreviations Used in the Notes

AGC—Alaska Game Commission

AK State Archives, RG#, SR#, VS#, folder#—Alaska State Archives, Juneau, Record Group, Series, Box, Folder.

Applegate Papers, Alaska State Library—Samuel Applegate Papers, 1892–1925, MS 3, Alaska State Historical Library, Juneau.

Ashbrook Papers, Smithsonian—F. G. Ashbrook Papers, circa 1915–1965, Record Unit 7143, Smithsonian Institution Archives, Washington, DC.

NARA—National Archives and Records Administration, Pacific Alaska Region, Anchorage.

Territorial Governors Correspondence—General Correspondence of the Alaskan Territorial Governor, 1908—1958, MFAR 27, Alaska State Historical Library, Juneau.

UAF Archives—University of Alaska Fairbanks Archives.

www.measuringworth.com—"Seven Ways to Compute the Relative Value of a U.S. Dollar Amount, 1774 to Present," www.measuringworth.com

Preface

1. The value of 1930 dollars in 2010 using the Consumer Price Index can be approximated by multiplying by 13.3. www.measuringworth.com (accessed 4-9-11).

Endnotes

Chapter 1

1. The Commander Islands (*Komandorskiye Ostrova*) are named for Commander Vitus Bering who died there on his way back from the discovery of Alaska. A geologic extension of the Aleutian chain, the islands are 200 miles closer to Kamchatka than to the tip of the Aleutians. They have been historically considered Siberian.

2. Coxe 1996, 44; Black 2004, 69.

3. Mai-mai-cheng housed only men because Chinese women were forbidden contact with foreigners. Chevigny 1965, 44–45.

4. At her 1762 coronation, Catherine the Great wore an embroidered velvet robe trimmed with sable and lined with hundreds of ermine skins. It was valued at 25,000 rubles (more than $155,000 in today's dollars; see endnote 5 for calculation). *The Romance of Furs* (Washington, DC: Woodward and Lothrop, 1936), 5.

5. The blue fox pelts were from the Commander Islands. Buskirk and Gipson 1980, 51–52. The current value of mid-eighteenth-century rubles is difficult to determine. By the mid-nineteenth century, 112,220 rubles would be worth $692,400 in twenty-first-century dollars. "Value of a Ruble in the 1860s," www.answers.google.com/answers (accessed 7-8-08). The furs were even more valuable at a point of trade. In 1746, a sea otter pelt worth 20 to 40 rubles in Kamchatka was worth 60 or 70 rubles at Kiakhta. Berkh 1974, 13 and 98.

6. Black 2004, 30; Berkh 1974, 7. Bering had also tried to find DeGama Land, which appeared in sketchy form on the Joseph N. Delisle chart he was using. Stejneger 1936, 249 and 257.

7. The government tax was 10 percent of the furs on an arriving ship. The tribute, *iasak*, was one pelt rendered to the Tsarina by each Aleut man who hunted under her "protection." Traders and government representatives assigned to some ships had difficulty explaining the benefits of *iasak* and persuading hunters to contribute pelts. Berkh 1974, 100–102.

8. Some *shitiki* were lashed together with strips of whale baleen. Others used reindeer-hide for sails to avoid the cost of importing sailcloth. Fisher and Johnson 1993, 99; Berkh 1974, 12–13 and 97.

9. Dmytryshyn and Crownhart-Vaughan and T. Vaughan 1988, 206. Crews were usually split between Russian *promyshlenniki* (Siberian hunters and townsmen) and Kamchadals (natives of Kamchatka). The former were said to be more enterprising and able defenders of a ship; the latter tougher and better able to withstand scurvy, thus cheaper to provision. Coxe 1996, 9.

10. Chevigny 1963, 39.

Endnotes

11. Black 1984, 79, n. 9. One crew-constructed boat, the *Kapiton*, cobbled together on Bering Island from two wrecks and driftwood, was 40 feet long. After returning to Kamchatka, the reconstituted *Kapiton* made another voyage during which she was definitively wrecked. Berkh 1974, 8–9, 18–20.
12. Black 2004, 69.
13. Today some taxonomists classify the arctic fox as *Vulpes lagopus*.
14. On the American mainland, homozygous recessive white foxes comprise 98 percent of the population. Heterozygotes may be any shade from dark blue to smoky white. Heterozygotes of intermediate "aluminum" shades would, two centuries later, be bred for the specialty fur market. Novak et al. 1987, 395. Prices listed are from the 1780s. Coxe 1996, 13 and 16.
15. Another mammal at risk, Steller's sea cow (*Hydrodamalis gigas*), hunted from the Commander Islands to provision Russian ships, was extinct by 1768.
16. Bailey 1993, 3.
17. Masterson and Brower 1948, 90.
18. Sowls, Hatch, and Lensink 1978, Map 13. iba.audubon.org (accessed 6-8-10).
19. Masterson and Brower 1948, 60–61.
20. Chevigny 1965, 36; Berkh 1974, 10. Coxe details the gifts to the Tunulgasen, but ascribes them to a later Tolstykh expedition. Coxe 1996, 7 and 44–45. Lydia Black offers a modern spelling for Tunulgasen, indicating that it was a title rather than a given name: "the Tukugassin of the Near Islands." Black 1984, 75.
21. Bailey 1993, 33.
22. Veniaminov 1984, 335–336. This type of trap and two others made of wood, stones, and sinew continued to be used by Aleuts along with steel traps through the 1930s. For a description of the traps, see Shade 1949, 41–44.
23. Black 2004, 76, n. 35.
24. Berkh 1974, 18.
25. In *V. vulpes*, formerly known as *V. fulva*, two genes ("A" dominant, "a" recessive and "B" dominant, "b" recessive) are paired so that a typical red fox can be represented as AABB, a cross fox as AaBB, a full silver fox as AAbb, a half-silver fox as aaBb, and a full black fox as aabb. *Yearbook of Agriculture* (Washington, DC: GPO, 1937). In subsequent years, the identification of animals with hair color mutations made it possible to develop genetic strains not found in the wild.
26. Novak et al. 1987, 379–380.
27. Coxe 1996, 16.
28. Black 2004, 69 and 76, n. 33.
29. Bancroft 1960, 253.

Endnotes

30. Kim MacQuarrie, "Bering Island," www.pbs.org/edens/kamchatka (accessed 6-18-10). Bancroft 1960, 191, n. 33.
31. D. Jones 1980, 37.
32. Elliott 1976, 8.
33. D. Jones 1980, 11; Smith 2001, 11. Pribylov's first name is rendered Gerasim by some authors. Igadik can also be spelled Igadagax. Amiq is thought to be St. Paul Island.
34. Veniaminov refers to "the preeminence of Pribylov in skillful seamanship among all the mariners of the time in the Aleutians." Veniaminov 1984, 135.
35. The Pribilofs also include Walrus Island and Otter Island—tiny protrusions of inactive underwater volcanoes.
36. In addition to the two winters on St. Paul and St. George, Pribylov and his crew hunted elsewhere before they returned to Kamchatka in 1789. Their furs included 6,794 blue fox pelts, but it is unknown how many came from the Pribilofs. Berkh 1974, 104. Pribylov last hunted at the islands in 1790. Pierce 1990, 412.
37. Smith 2001, 36.
38. Authorities give varying estimates. Osgood and Preble and Parker 1915, 106; Smith 2001, 62.
39. Bailey 1993, 5. "Lesser" apparently referred to small, uninhabited, western islands.
40. Black 2004, 76 and 199.
41. Khlebnikov 1994, 158.
42. Tikhmenev 1978, 359.
43. Black 2004, 90 and 148. One author suggests he was disabled by alcoholism. Torrey 1983, 49.
44. In 1764, Tolstykh's ship, *Andreian i Natalia*, wrecked on rocks near an Aleutian Island. The crew was able to rebuild the ship, but as they approached home, they again wrecked off Kamchatka Cape. Although all the crew and cargo survived, the ship was too far gone to rebuild.
45. Fifty-nine men out of a crew of sixty-three perished with Tolstykh. Oleksa 1992, 85.
46. Berkh 1974, 28.
47. Chevigny 1965, 36–37; Berkh 1974, 38; Pierce 1990, 510.
48. Black 1984, 157. Early in the American period (1880), Agattu Aleuts continued to press this concern. Bailey 1993, 35.
49. alaskamaritime.fws.gov (accessed 5-25-08).

Endnotes

50. Buskirk and Gipson 1980, 49–50; Bailey 1993, 5–6, 13–14, 19, 37; Evermann 1914, 29–30; Black 1984, 102. An attempt to transplant ground squirrels to Attu Island in the late 1840s failed. Tikhmenev 1978, 359.
51. Dall 1870, 498–499.

Chapter 2

1. Washburn chose not to use Martin, his given name.
2. Examples include Niedieck 1909, 162; and "The Blue Fox Industry in Alaska," *The Fur Farmer Magazine* 4, no. 7 (February 1928): 14.
3. Washburn 1901, 357–378.
4. Various sources date the founding of the Semidi Company from 1882 to 1885. Taylor, an early Treasury agent in the Pribilofs, was president. Morgan and Redpath had worked for the Alaska Commercial Company on St. Paul Island. Washburn, vice president, was apparently the only partner who had not been to the Pribilof Islands.
5. The stated purpose of this first wildlife refuge in the United States was to protect the seals, to create a government protectorate for the islands' Aleuts, and to bring money into the Treasury by leasing seal harvesting rights to a single company that would render rent and royalties. D. Jones 1980, 15.
6. D. Jones 1980, 1. Congress had previously voted for the Louisiana Purchase based on its presumed wealth of furs, and the nation had received good economic return. The government had similar expectations for Alaska.
7. Crompton 1920.
8. "Transcript of Notes from St. Paul Island, 1872–1909," compiled by Walter L. Hahn, 1910–1911, p. 94, Series 91, Box 2, Folder 7, Federal Records Collection, U.S. Fish and Wildlife Service, Division of Alaska Fisheries, Record Group 22, MS 51, Alaska State Library, Juneau.
9. M. Kutchin, "Fox Breeding on the Alaskan Islands," *Scientific American* 82 (April 1900): 242. Based on the Consumer Price Index, the value of $10 in 1895 would equal about $270 in 2010. Throughout this chapter, multiplying quoted dollars by 25 will approximate 2010 values. www.measuringworth.com (accessed 4-9-11).
10. 55th Congress, Sess. II, Ch. 288, May 14, 1898. 57th Congress, Sess. II, Ch 1002, March 3, 1903. See also revised statutes, secs. 2291, 2292, 2305. Fur farming apparently did not satisfy the Homestead Act's goal of bringing wilderness under cultivation and supplying food for the nation.
11. An agent of the General Land Office stated that if there were no leases, when "enterprising outsiders try to land on [a fox] island, the mortality rate of Alaskan

islands is apt to rise. . . . present tenants need to be given more formal protection than the fortunes of war with a double barrel shotgun." "Fox Raising in Alaska," *Daily Alaska Dispatch*, January 4, 1904.

12. The right of Alaska Natives to all land currently "in their use or occupation" was guaranteed by the Organic Act of 1884, but their means for securing this land was vague.
13. "NOTICE," Box 2, file 8, MS 3, Applegate Papers, Alaska State Library.
14. Tingle 1897, 33–34.
15. Captain F. F. Feeney had made an attempt to establish a fox farm 1880 on Long Island with two pairs of black foxes, but they were poached during the first winter. He tried again in 1889 but continued to lose animals. When the Semidi Company bought his house, sheds, and stock for $8,000 in 1895, he had forty-five cattle, some sheep, and eight black foxes, which the Semidi partners replaced with blues. Huston 1963, 19.
16. Janson ca. 1985, chapter 3, pp. 1–2.
17. Editorial note appended to Washburn's chapter. See Washburn 1901, 365. The up-to-date launch ran on boiling petrol; the vapor powered an engine similar to a steam engine.
18. Willoughby 1925. *Rocking Moon*, motion picture, directed by George Melford (Metropolitan Picture Corp. of California, 1926).
19. *Fur Trade Review* 26 (August 1, 1898): 343.
20. Washburn 1901, 364–365; Osgood 1908, 21.
21. Johnston, ca. 1940, 38; J. Forester and A. Forester 1980, 41–42.
22. Osgood 1908, 5.
23. Washburn 1901, 365; Bailey 1993, 9.
24. Osgood 1908, 107.
25. McAllister 1943, 10–11.
26. This treaty, established by the Tribunal of Fur Seal Arbitration, failed to successfully regulate pelagic hunting. In 1912, it was replaced by a more enforceable treaty that resulted in a rise in fur seal numbers.
27. Judge 1913, 60–69; Osgood 1908, 107; Preble and McAtee 1923, 1 and 104.
28. Huston 1963, 22. M. Kutchin, "Fox Breeding on the Alaskan Islands," *Scientific American* 82 (April 1900): 242.
29. Washburn 1901, 360. The Semidi Company farmed North and South Semidi, Long, Whale, Chirikof, and Marmot Islands. They sold stock to farmers on Pearl, Little Naked, Goose, Green, Demidof, Deranof, and Ugak Islands. Washburn commented that Attu did not have to be stocked, as it had wild blue foxes.

30. Janson ca. 1985, chapter 8, p. 1. Only the U.S. government on the Pribilofs sold more fox pelts in 1903, marketing 521. The six islands of the Semidi Company produced only 115 pelts. Osgood 1908, 107; Huston 1963, 23.
31. Janson ca. 1985, chapter 8, pp. 1–2; chapter 7, p. 5; chapter 9, pp. 1–2. The Kachemak Bay farmer was U. S. Ritchie.
32. S. C. Eby, "A Fox Farm on an Island of the Sea." *Alaska-Yukon Magazine* 12, no. 1 (February 1912): 7–10.
33. Underwood 1913, 283.
34. The Alutiiq name for the island, Achaka (Ucuaq), was said to mean "harborless." Margery Pritchard Parker, "A Northern Crusoe's Island," *National Geographic*, 44, no. 3 (September 1923): 313.
35. In 1893, Middleton Island was the site of gold excitement with a number of claims staked but soon abandoned.
36. Janson ca. 1985, chapter 5, pp. 1–3. S. C. Eby, "A Fox Farm on an Island of the Sea." *Alaska-Yukon Magazine* 12, no. 1 (February 1912): 7–10. Underwood 1913, 284–290. Smith may have had good reason for suspecting Japanese poachers. In 1910, the sealing schooner *Tokio Maru* was seized by a U.S. revenue cutter for poaching foxes in the Shumagin Islands. "Jap Sealer Seized by the Cutter Tahoma," *Seward Weekly Gateway*, July 16, 1910. Not long after moving to Washington, Smith encountered one of civilization's communicable diseases and died of typhoid. Underwood 1913, 290.
37. Stuck 1917, 281.
38. E. Jones 1915, 122.
39. Ned Dearborn, "Report on Fur Farming in Alaska, Summer 1915," p. 8, Series 10, Box 15, Folder 8, Ashbrook Papers, Smithsonian.
40. Ned Dearborn, "Report on Fur Farming in Alaska, Summer 1915," pp. 8–9, Series 10, Box 15, Folder 8, Ashbrook Papers, Smithsonian.
41. Huston 1963, 15.
42. Henshaw 1912, 659; Ashbrook 1923, 4.
43. By 1922, there would be 500 silver fox farms in the northern states. Ashbrook 1923, 6.
44. Janson ca. 1985, chapter 3, p. 1.
45. Washburn 1901, 363–364.
46. Bailey 1993, 18–21, 23.
47. Osgood 1908, 22. M. Kutchin, "Fox Breeding on the Alaskan Islands," *Scientific American* 82 (April 1900): 242. Evermann 1914, 26.
48. The dimensions are from a report describing a Haines farm that housed thirty "tame and playful" silver foxes. Evermann 1914, 22.

Endnotes

49. Stuck 1917, 281. *Annual Report of the Commissioner of Fisheries to the Secretary of Commerce, 1918* (Washington, D.C.: GPO, 1920), 71. "Chief Titus" was probably Titus Alexander of Hot Springs.
50. Ned Dearborn, "Report on Fur Farming in Alaska, Summer 1915," p. 4, Series 10, Box 15, Folder 8, Ashbrook Papers, Smithsonian.
51. Ned Dearborn, "Report on Fur Farming in Alaska, Summer 1915," p. 16, Series 10, Box 15, Folder 8, Ashbrook Papers, Smithsonian.
52. House Committee on Expenditures in the Department of Commerce and Labor, *Hearings on H.R. Res. 73 to Investigate the Fur Seal Industry of Alaska* (Washington, DC: GPO, 1911), 340–342.
53. Heideman 1909, 14–15 and 31.
54. Heideman 1910, 17–18.
55. Letter from the Alaska Fur and Silver Fox Company to unnamed shareholders, 1916. Letter glued inside a pamphlet, "The Story of a Fox"—a prospectus for the Alaska Silver Fox and Fur Farms Company." The pamphlet was published by a Fairbanks company with a name similar to that used by Heideman and partners. Wickersham Collection, item number 3239, Alaska State Library, Juneau. A difference between the two companies was that the Fairbanks company, managed by J. S. Sterling, actually shipped 65 live silver fox breeders to Seattle in 1915. "Foxes There," *The Alaska Citizen*, October 11, 1915.
56. Bower and Aller 1918, 110.

Chapter Three

1. "The Forest Reserve," *Ketchikan Mining Journal*, November 8, 1902. "Forest Reserve Set Aside," *Seward Weekly Gateway*, August 3, 1907.
2. Pinchot 1910, 42.
3. Miller 2001, 155. Pinchot is widely credited for this definition of conservation, although it may have been adapted from English philosopher Jeremy Bentham.
4. The legal distinction between a permit and a lease was explained in a government memo. A *permit* is revocable at any time; a *lease* runs for a specific term. Memorandum from Acting Solicitor for Department of Agriculture [signature unreadable] to Chief, Biological Survey, 27 December 1918. For southeast Alaska fur farmers, the distinction seemed unimportant as permits were usually issued for a stated term and, like leases, revoked only for cause—usually failure to farm. In the Aleutians just prior to WWII, the limits of a permit would become an important issue.

Endnotes

5. The value of $25 in 1910 is equivalent to about $590 in 2010 dollars. For the remainder of this chapter, the dollar amounts quoted can be multiplied by 24 to get the approximate value in 2010 dollars, according to the Consumer Price Index calculation at www.measuringworth.com (accessed 4-9-11).

6. U.S. Forest Service, Department of Agriculture, Land Case Files 1910–1974, Record Group 95, NARA.

7. The Division of Economic Ornithology created in 1886 became the Division of Biological Survey in 1896 and was renamed The Bureau of Biological Survey in 1905. Merriam, under pressure and at odds with Department of Agriculture goals, resigned his post in 1910 after 25 years of service. Keir B. Sterling, "Builders of the U.S. Biological Survey, 1885–1930," *Journal of Forest History* 33 (October 1989): 180–187.

8. Osgood 1908; Cameron 1929, 223–251. This section of Cameron's book lists all Biological Survey publications prior to 1929.

9. Harding 1909.

10. Huston 1963, 54; Ned Dearborn, "Report on Fur Farming in Alaska, Summer, 1915," p. 12, Series 10, Box 15, Folder 8, Ashbrook Papers, Smithsonian.

11. *Annual Report of the Governor of Alaska to the Secretary of the Interior, 1915* (Washington, DC: GPO, 1915), 36. Subsequent fur prices in this chapter are from the same source.

12. The farmer, "Shorty" Hunter, captured a female marten one summer. It lived in his house along with his tomcat and produced a litter the following spring. "New Fur Possibility," *The Fur Farmer Magazine* 1, no. 3 (April 1923): 7. Marten reproduction would be scientifically outlined in the mid-1930s when scientists at the U.S. Fur Animal Experiment Station in New York proved that that although marten mate in early summer, the fertilized egg does not implant in the uterus and begin to grow until the following winter.

13. *Annual Report of the Commissioner of Fisheries to the Secretary of Commerce for the Fiscal Year 1917* (Washington, DC: GPO, 1919), 59. J. Underwood 1913, 284; Janson ca. 1985, chapter 16, pp. 1–2; Huston 1963, 132; Bower and Aller 1916, 116.

14. E. Jones 1915, 124; Evermann 1914, 73–74.

15. Dearborn 1916, 490–491.

16. "Rush Starts Skunk Farm," *Skagway Daily Alaskan*, February 5, 1904. In this headline, "rush" refers to the rapid expansion of the fur farms before WWI, which would be eclipsed by a larger stampede in the 1920s.

17. M. Kutchin. "Fox Breeding on the Alaskan Islands," *Scientific American* 82 (April 1900): 242.

Endnotes

18. Bear and silver fox prices from "The Fur Market," *The Alaska Appeal*, September 30, 1897. The blue fox price, which is lower than the $50 prime price quoted by M. L. Washburn, is based on the actual 1901 sales of blue fox skins from Samalga Island. Huston 1963, 29. Examples of Alaska "bear men" include Stan Price of Pack Creek on Admiralty Island, Charlie Vandergaw in the Susitna Valley, and an entire gold dredge crew on the Hogatza River in the Koyukuk Valley.
19. See endnote 23 below.
20. Executive Order 1733, March 3, 1913.
21. Unalaska City School District 1986, 196–198. Applegate was also author of the Aleutian section of the 1890 U.S. census.
22. Treasury agents did not reveal that the fox population had slumped, and Applegate's repeated inquiries about the delay received little response. See correspondence between Applegate and Treasury Agent J. B. Crowley in Box 2, File 6, Applegate Papers, Alaska State Library.
23. Assistant Secretary of the Treasury to Applegate, August 21, 1900, Box 2, File 6, Applegate Papers, Alaska State Library. Applegate's papers reveal the bureaucratic problems in the Aleutians. In 1904, the Commerce Department took over the management from the Treasury Department. In 1913, this shifted to joint responsibility by Commerce (in charge of harvesting farmed foxes) and by Agriculture (in charge of propagating foxes). When Applegate asked to expand his farm to nearby islands, the Department of Agriculture wrote that it refused the request. The Department of Commerce wrote that it had no objection. Both letters were dated July 7, 1913. Applegate expanded his farm. W. F. Bancroft to Applegate, 7 July 1913; and H. M. Smith to Applegate, 7 July 1913, Box 2, File 8, Applegate Papers, Alaska State Library. In 1920, the Bureau of Biological Survey became the sole leasing agent for government islands in the Aleutians and near Kodiak.
24. Applegate's name and that of his boat, *Nellie Juan*, now label six geographic features in the Aleutians and Prince William Sound.
25. Unalaska City School District 1986, 196.
26. Applegate to Secretary of Commerce, 15 July 1914, Box 2, File 8, Applegate Papers, Alaska State Library.
27. For example, he was proud of his acumen in saving $300 a year by employing two Aleuts to manage his island in place of his previous white manager whose salary had been $540 per year. Applegate to Secretary of Commerce, 15 July 1914. He also asked the government to respect his white children and "hold them at least a little more worth" than Aleut children. Applegate to Secretary of Agriculture, 6 May 1915, Box 2, File 8, Applegate Papers, Alaska State Library.
28. Applegate to Secretary of Agriculture, 6 May 1915, Box 2, File 8, Applegate Papers, Alaska State Library.

Endnotes

29. Applegate to A. K. Fisher, 7 March 1916, Box 2, File 8, Applegate Papers, Alaska State Library. Applegate was unaware that, as described later in this chapter, government-supplied food to Pribilof Aleuts depended on profits from fox pelts.

30. Nicholas Bolshanin to D. F. Houston, Secretary of Agriculture, 5 May 1916. Bolshanin's correspondence in this chapter taken from copies in the research files of Sarah McGowan, author of "Fox Farming: A History of the Industry in the American Period," in Black 1999, 244–251. Bolshanin's given name was Nicholai, but he more frequently used the Anglicized version, Nicholas.

31. Bolshanin to Houston, 25 August 1916.

32. Bolshanin to E. W. Nelson, Chief, Bureau of Biological Survey, 24 August 1919. Applegate's suggestion that new farmers be allotted foxes from the Pribilofs had been rejected, and the price of breeding pairs sold by well-established white farmers was out of Aleut reach.

33. Morgan 1992, 122.

34. "Memo—Bolshanin," unsigned, dated Spring 1916, Box 2, File 4, Applegate Papers, Alaska State Library.

35. *Annual Report of the Chief of the Bureau of Biological Survey, 1921* (Washington DC: GPO, 1921), 32.

36. Before their final move, the Applegate family was already spending winters in California, where Samuel Applegate died in 1925. Nicholas Bolshanin continued fox farming until 1940 as the absentee permit holder on two small Aleutian Islands (Ulak and Kavalga) near the Alaska Peninsula. He lived in Sitka until the year before his death in Seattle in 1957.

37. Evermann 1913, 73–74. Ned Dearborn, "Report on Fur Farming in Alaska, Summer, 1915," p. 12, Series 10, Box 15, Folder 8, Ashbrook Papers, Smithsonian. Dearborn doubted that eagles were a serious threat after he examined an eagle nest and found only one fox skull amid a mass of fish bones.

38. In 1915, the Alaska legislature had already enacted a $10 bounty on wolves to protect moose, sheep, and caribou desired by human hunters. In the States, animals subject to historic bounties have included bobcats, lynx, ground squirrels, porcupines, tree squirrels, and foxes.

39. The Audubon Ornithologists Union, which had previously gone on record opposing bounties on predatory birds, debated for three years before deciding not to publicly oppose Alaska's law. Barrow 1998, 47. The U.S. Commissioner of Fisheries saw in the bounty law a direct economic benefit to fur farmers: Eagle bodies could be used as fox food. *Annual Report of the Commissioner of Fisheries to the Secretary of Commerce for the Fiscal Year, 1917* (Washington, DC: GPO, 1918), 66. Alaska's eagle bounty law underwent several changes, and was

eliminated for the final time in 1953. Estimates of the number of eagles killed for bounty vary from a low of 128,000 to a high of 140,000.

40. J. Forester and A. Forester 1980, 24.
41. Evermann 1914, 24; Huston 1963, 21–23.
42. Huston 1963, 21.
43. Osgood 1908, 17–19.
44. W. J. Erskine, "Report of the Kodiak Fox Farm, December 1, 1915," pp. 3–4, Series 7, Box 9, Folder 6, Ashbrook Papers, Smithsonian.
45. Andrew Grosvold, "Blue Fox Ranching on the Westward Islands," *The Pathfinder* 3, no. 7 (June 1922): 5.
46. "Easy to Raise, Hard to Catch," *The Tacoma Times*, December 12, 1911.
47. Ned Dearborn, "Report on Fur Farming in Alaska, Summer 1915," p. 9, Series 10, Box 15, Folder 8, Ashbrook Papers, Smithsonian.
48. During WWI, the Bureau of Fisheries would sound a warning that new farmers paying $300 to $500 for a pair of silver foxes with the intent to raise breeders would experience a "keenly-felt fall in earnings" when supply exceeded demand. *Fox Farming in Alaska*, Bureau of Fisheries, Department of Commerce. Document 834, repr. in *Fur News* (June 1917): 16.
49. In 1911, Northern Navigation Company also memorialized Washburn by launching the *M. L. Washburn*, a 120-foot river steamer built in St. Michael for the Yukon River trade.
50. W. J. (Wilbur Julian) "Billy" Erskine's grandfather was Melville C. Erskine; his father was Wilbur F. Erskine.
51. Tom L. Alton, "W. J. Erskine: Early Alaska Merchantilist," manuscript, n.d., pp. 2–3, Biography File: W. J. Erskine, Alaska State Library, Juneau.
52. Erskine's partners were Nicholas Gray, former manager of Alaska Commercial Co. in Unalaska; Peter Blodgett, former operator of the Semidi fox farm on Chirikof Island; and Karl Armstrong, Kodiak Deputy Marshal and owner of a fox farm in Kalsin Bay. Blodgett was part owner of the cannery that supplied Long Island with fish heads. The chief caretaker, Durrell Finch, former manager of Alaska Commercial Co. at Tyonek, would eventually start a fox farm near Unalaska and operate it until 1940.
53. Erskine, "Report of the Kodiak Fox Farm, December 1, 1915," bound typescript, Series 7, Box 9, Folder 6, Ashbrook Papers, Smithsonian. Erskine recommended that the government eliminate co-management of fur farms by the Departments of Agriculture and Commerce, that it require permits for capturing wild breeders, that it ban steel traps for catching breeders, that it forbid

shipment of live wild foxes out of Alaska, and that it provide means for fur farmers to gain title to their land.

54. The market for breeders actually remained strong into the mid-1920s when even blue fox pairs sold for $250 to $400. Ashbrook to E. W. Nelson, 5 February 1923, Series 10, Box 16, Folder 7, Ashbrook Papers, Smithsonian.

55. The eruption was not Mt. Katmai itself but a neighboring newborn mountain, Novarupta, on the Alaska Peninsula. Erskine was one of the few people who personally experienced the Nome gold rush, the San Francisco earthquake, and the Katmai eruption.

56. William C. Redfield to Erskine, 4 April 1915, Series 7, Box 9, Folder 6, Ashbrook Papers, Smithsonian.

57. Bower and Aller 1918 (report for 1916), 137. Another reason for this relaxed attitude may have been ambiguity about the law. An early Treasury agent counted 22 fox farms, half of which were operated "without any embarrassing pecuniary formalities," but opined that the Homestead Act exclusion, which referred to land "leased or occupied" for the purpose of fox farming, legitimized squatters by including the word *occupied*. H. M. Kutchin 1898, 32.

58. Bower 1920 (report for 1918), Appendix VII, 72.

59. Alaska Commercial Company held the Pribilof fur contract from 1870 to 1890; Northern Commercial Company held it from 1891 to 1910. American conservationists defeated these companies, which they suspected of cheating on the fur seal treaty, and persuaded Congress to end all commercial leases.

60. In addition, the government had already realized a significant savings by eliminating the commercial middleman. Alice Lloyd-Jones 1918, 26. The St. Louis outlet, which later changed its name to Fouke Fur Company, continued to receive Pribilof pelts until its contract was terminated in 1961. D. Jones 1980, 51–52.

61. Applegate to A. K. Fisher, 7 March 1916, Box 2, File 8, Applegate Papers, Alaska State Library.

62. D. Jones 1980, 67–72; Torrey 1983, 115–117 and 138–139.

63. *The Romance of Furs* (Washington, DC: Woodward and Lothrop, 1936), 8.

64. Per capita, more Alaskans served in World War I than citizens of any of the states. Greely 1925, 238.

65. Crosby 2003, 241.

66. Sale of breed stock did not stop entirely; in 1915, a Fairbanks farm shipped 65 live silver and blue foxes to a farm in Plattsburg, NY, at one-third the pre-war price. "Fur Farm Cuts Price of Skins," *The Tacoma Times*, October 4, 1915.

67. Bower and Aller 1919 (report for 1917), 106.

68. Bower and Aller 1917 (report for 1915), 109.
69. J. Forester and A. Forester 1980, 122. E. Jones 1915 (report for 1914), 118.

Chapter 4

1. In 1920, $17,000 was equivalent to $185,000 in 2010 by Consumer Price Index calculations. A 1920 dollar can be multiplied by 12.7 to approximate its 2010 value. www.measuringworth.com (accessed 4-9-11).
2. *Annual Report of Commissioner of Fisheries to the Secretary of Commerce for the Fiscal Year 1919* (Washington, DC: GPO, 1920), 68.
3. "100 Blue Fox Bring $10,000," *The Valdez Miner*, February 1, 1919.
4. Josephine Sather, "Fox Farm at Nuka Bay," Part 1: "The Island," *The Alaska Sportsman* 12, no. 7 (July 1946): 6.
5. Janson ca. 1985, chapter 8, p. 4.
6. Average fur prices from 1919 to 1924 in this chapter are from annual press releases by the Bureau of Biological Survey titled "Shipments of Furs from Alaska for the Year [with specific year stated]."
7. Crosby 2003, 241–256.
8. Ernest Thompson Seton, "Raising Fur-Bearing Animals for Profit," *Country Life in America* 9 (January 1906): 294. Seton dabbled in fox farming and illustrated Biological Survey bulletins. He also wrote an anthropomorphic tale that portrayed "the beautiful monogamy of the better-class fox." Seton 1909, Introduction.
9. *Annual Report of the Chief of the Biological Survey, 1921* (Washington, DC: GPO), 33.
10. A 1923 list identifies 189 blue fox island farmers in Alaska. U.S. Bureau of Biological Survey, "Island Blue Fox Ranchers of Alaska: Stocked January 1, 1923," AK State Archives, RG 106, SR 102, VS 1, folder "Island Information,"
11. *Annual Report of the Secretary of the Interior for FY 1922* (Washington, DC: GPO, 1922), 44.
12. "Fur Industry on Increase in S.E. Alaska," *The Alaska Weekly*, March 16, 1923. "Blue Fox Industry Expanding," *The Alaska Weekly*, April 6, 1923.
13. Margery Pritchard Parker, "A Northern Crusoe's Island," *National Geographic* 44, no. 3 (September 1923): 313. The Crusoes of the article appear to be a blend of the Ibachs and their subsequent caretakers, Mr. and Mrs. Fred Schiller.
14. An Indian village and fort had also once been located on the island. Goldschmidt and Haas 1998, 56.

15. Barrymore was an actor; Lippincott and Beach were authors. Beach describes visiting Joe and Muz Ibach at Lemesurier Island in his autobiography, *Personal Exposures*. Beach 1940. In his novel *Valley of Thunder*, a character, "Gus Brown," is modeled on Ibach. Beach 1939.

16. Don [sic] Cadzow, "Fur Farming above the Arctic Circle," *Fur-Fish-Game* (March 1917): 16–17.

17. "The Groom Comes out of the North," *Oswego Daily Palladium*, September 11, 1916. "Auburn Man in Alaska Sends Letter by Air," *Oswego Daily Palladium*, October 25, 1920.

18. Huston 1963, 78 and 80.

19. "Foxes Exported from Alaska," *Fairbanks Daily News-Miner*, September 19, 1924; "Fox Farming Spreads in Alaska," *Fairbanks Daily News-Miner*, April 2, 1925. Goshaw started farming white arctic foxes in 1922. He bred and re-bred white with blue foxes, seeking a "platinum"-colored fur. "Goshaw Talks on Fur Farms," *Nome Nugget*, March 22, 1930.

20. "Eskimos are Making Success of Fox Farming," *Fairbanks Daily News-Miner*, October 1, 1925.

21. In 1908, Hegness, driving his mail-freighting team, won Nome's 400-mile All-Alaska Sweepstakes in 119 hours. The last year of this classic race was 1917.

22. "On the Edge of the Arctic Ice," *The Fur Farmer Magazine* 4, no. 2 (August 1927): 27. Northern Whaling had a trading post at Cape Halkett where Hegness kept his foxes.

23. In 1928, Alaskan white fox pelts averaged $46. AGC, "Furs Shipped During the Year 1928." Ashbrook 1925, 33. In 1923, Harriet Rossiter was behind the times when she wrote that American women preferred blue fox skins because "their genuineness is guaranteed by their color which no dye can reproduce." Harriet Rossiter, "Alaska's Fur Industry, Its Decline and Rise," *Fur Trade Review* (January 1923): 174.

24. Some survived long enough to be noted in government reports—e.g., mice on Chankliut Island (1919) and ground squirrels on Nohatamie (1920). Bailey 1993, 37 and 52–53.

25. *Annual Report of the Chief of the Biological Survey, 1921* (Washington, DC: GPO, 1921), 33. *Annual Report of the Secretary of the Interior for FY 1922* (Washington, DC: GPO, 1922), 42. "Huge Profits in Fox Farms," *The Nenana Daily News*, June 21, 1923.

26. Alaska Territorial Department of Audit, Fox Brand Program (file collection), AK State Archives, RG 106, SR 102, Box 3, Vol. 2, folder 3 and Vol. 3, folders 1, 2. Schlung and Nikolski students 2003, 39.

27. "Muskrats Make One Grubstake," *Fairbanks Daily News-Miner*, April 6, 1923.
28. Janson ca. 1985, chapter 16, pp. 3–4. When Zimmerman stocked Brothers Island with raccoons and skunks in 1919, he reported that there were a few skunk and raccoon descendents from a previous farmer's stocking in 1913. Bower 1919, 68.
29. Marten and beaver were especially valued because the Alaska Game Commission intermittently closed trapping for these species to save wild stock. Unfortunately for farmers, marten were hard to breed and beaver hard to contain. AGC, "Report of the Executive Officer to the Alaska Game Commission for the period January 1, 1940 to December 31, 1940," MF 56, Alaska State Library, Juneau.
30. "$10,000 Mink Coats," *Popular Science* (November 1935): 36–37.
31. AGC, "Annual Report of the Alaska Game Commission to the Secretary of Agriculture, 1925," Box 6, AGC Collection, UAF Archives. AGC, "Annual Report of the Alaska Game Commission to the Secretary of Agriculture, 1930," Box 1, folder 4, AGC Collection, UAF Archives.
32. Chapter 42, Alaska Session Laws of 1921.
33. Frank A. Jones to editor, *The Nenana Daily News*, August 11, 1921.
34. Chapter 20, Alaska Session Laws of 1927. The law had already been amended in 1923 in an unsuccessful attempt to solve collection problems. Chapter 89, Alaska Session Laws of 1923.
35. Advertisements, some abridged, from *Fur Farmers' Bulletin: Official Organ of the Southeastern Alaska Blue Fox Farmers Association*, December 14, 1923; January 16, 1924; March 20, 1924; May 31, 1924.
36. "Successful Fox Farmer," *The Nenana Daily News*, May 17, 1923.
37. *The Fur Farmer Magazine* 1, no. 8 (November 1923): 6.
38. *The Fur Farmer Magazine* 3, no. 4 (July 1925): 30.
39. *The Fur Farmer Magazine* 3, no. 1 (April 1925): 18. *The Fur Farmer Magazine* 1, no. 7 (October 1923): 6. Advertisement, Anchor Point Silver Fox Farm, *Fairbanks Daily News-Miner*, May 14, 1924.
40. "Charles Darwin Garfield," obituary *Seattle Times*, September 25, 1961, p. 37. Garfield's publication is described more fully in Chapter 6, endnote 9.
41. *Fairbanks Daily News-Miner*, February 27, 1920. *The Alaska Dispatch*, December 22, 1922. *The Nenana Daily News*, June 21, 1923.
42. "Phoney Furs," *The Fur Farmer Magazine* 1, no. 7 (October 1923): 17. S. V. B. Miller, "Why Cats Are Wild," *The Fur Farmer Magazine* 1, no. 9 (December 1923): 10. "Inspection That Does Not Inspect," *The Fur Farmer Magazine* 3, no. 5 (August–September 1925): 32. Large-scale scams did occasionally make

it to the courtroom. In 1931, G. R. Flynn promised a California court to repay $56,000 to investors who bought fake Alaskan fur farm stock. The reimbursement was to come out of proceeds from his "oil and timber lands in Alaska." He received five years' probation. "Sells Fake Stock Alaska Fur Farms," *Fairbanks Daily News-Miner*, January 16, 1931.

43. "Coast Guard Cutter Will Protect Fox Farmers," *The Fur Farmer Magazine* 1, no. 3 (April 1923): 7.
44. Susie Pederson, "Ed Ophiem, Sr.: Fox Farming" *Elwani* 6, (1979): 6. "The Poacher Problem," *The Fur Farmer Magazine* 1, no. 1 (February 1923): 4.
45. Chapter 14, Alaska Session Laws of 1925.
46. "Would Brand Foxes for Protection," *The Pathfinder* 3, no. 1 (November 1921): 11.
47. Alaska Territorial Department of Audit, Fox Brand Program Rules, AK State Archives, RG 106, SR 102, Box 1, folder "Fox Brand Inquiries."
48. "Undeclared Fox Skins Seized in Seattle," *Wrangell Sentinel*, January 14, 1930.
49. "Alleged Poacher Killed," *Fur Farmers' Bulletin* 1, no. 4 (March 20, 1924): 1–3.
50. Ashbrook started with the U.S. government in 1914 as a junior husbandman. By 1924, he had advanced to head the Division of Fur Resources, which oversaw fur farms and wild fur harvests. In 1949, he left the fur farm industry behind and shifted to the U.S. Fish and Wildlife Service to head Wild Fur Animal Investigations. Ashbrook retired in 1957.
51. Alaska Territorial Department of Audit, Fox Brand Program Rules, AK State Archives, RG 106, SR 102, Box 1, folders "Island Information" and "Fur Farm Locations." "Expert Blue Fox Farmer," *Fur Farmer Magazine* 2, no. 12 (March 1925): 22. "Mrs. Williamson, Williamson Silver Fox Ranch Kusilof," *Fur Farmer Magazine* 2, no. 3 (March 1924): 16. "Future of Fox Farming," *The Alaska Weekly*, June 5, 1925. One of the inspectors, Perry A. Cole, remained in Alaska to develop his own 800-acre fur farm on the Kenai Peninsula. "Cole Plans a Big Scale Fox Industry," *The Alaska Weekly*, April 27, 1928.
52. Josephine Sather, "Fox Farm at Nuka Bay," four-part series in *The Alaska Sportsman*, 1946. C. Jones 2006, 66–68 and 73. Josephine was widowed on Nuka Island but continued the fox farm with her third husband, Pete Sather. Her farm made it through the Depression, selling $8,840 worth of pelts in 1939, but ended in the early 1950s when prices failed to recover after WWII. "Rich Fur Pelting Done by the Sathers. Biggest of All Years," *Seward Gateway*, December 30, 1939.
53. "Heiress Plans Raising Foxes Westward Isle." *Juneau Empire*, November 30, 1937. "Voluntary Exile," *The Alaska Sportsman*, November 1939, 8–9, 30–33.

54. Examples include Hilda Harbin Taylor (Taylor 1994) and Lillian Walker, interviewed by Debbie Turner and Karen Brewster, September 2, 2003, University of Alaska Oral History Project H2002-10-09, Part I.
55. Willoughby 1925. *Rocking Moon*, directed by George Melford, Metropolitan Picture Corporation of California, 1926. Barrett Willoughby to W. J. Erskine, 19 November 1925, Wilson Fiske Erskine Collection, 1884–1964, Box 2, folder 3, UAF Archives.
56. Osgood 1908, 20.
57. "Make Foxes Pets," *The Pathfinder of Alaska* 3, no. 13 (November 1922): 17. [Volume numbering for 1922 contains errors; this issue should be labeled volume 4, no. 1.] C. E. Zimmerman, "Domesticating Alaska Blue Foxes," *The Pathfinder of Alaska*, 4, no. 11 (September 1923): 1.
58. F. Berry, "Fur—Our Great Future Industry," *The Pathfinder of Alaska* 4, no. 8 (June 1923): 2.
59. "Pelts from the Farmers," *The Fur Farmer Magazine* 1, no. 7 (October 1923): 12.
60. "Berry to Engage in Fur Farming," *Fairbanks Daily News-Miner*, November 21, 1927.
61. Lyudmila N. Trut, "Early Canid Domestication: The Farm-Fox Experiment," *American Scientist* 87 (March–April, 1999): 160–169.
62. C. E. Crompton, "Methods Employed in the Management of Blue Foxes," Report to the Commissioner of Fisheries, October 23, 1920. Box 2, folder 5, U.S. Fish and Wildlife Collection, MS 51 Alaska State Historical Library, Juneau.
63. Frank G. Ashbrook to Chief, Division of Wildlife Research, memorandum, 26 July 1946, Series 7, Box 8, folder 6, Ashbrook Papers, Smithsonian. Fox farming on St. George was consistently more efficient and productive than on St. Paul, but reported sales were usually combined.
64. Frank G. Ashbrook to E.W. Nelson, Chief, Bureau of Biological Survey, memorandum, December 19, 1922, 17–18, Series 10, Box 16, folder 8, Ashbrook Papers, Smithsonian. "Directors Meeting," *Fur Farmers' Bulletin* 1, no. 2 (January 1924): 4. One complaint noted that if Pribilof fox pelts had been taxed for the years from 1919 to 1921, the Bureau of Fisheries would have owed the territorial government $3,750. Janson ca. 1985, chapter 2, p. 6.
65. AGC, "Furs Shipped from Alaska December 1, 1919 to November 30, 1920." AGC, "Furs Shipped from Alaska During the Year 1925."
66. In 1923, when U.S. silver fox farmers sold 6,000 pelts, only 260 came from Alaska. "The Comeback of the Fur Bearers," *Popular Mechanics*, 76, no. 5 (November 1941): 94. AGC "Furs Shipped from Alaska, December 1, 1922 to November 30, 1923."

Endnotes

Chapter 5

1. Rakestraw 1994, 125. The Juneau District Forester was Charles H. Flory, author of a 1924 letter to the Washington, DC, office that stated: "Where the only Indian claims are abandoned gardens or shacks we have not regarded them as legal rights either on fox islands or elsewhere. . . . I am of the opinion that we have been rather too liberal in . . . [requiring] applicants for fur farm leases to secure clearance from Indians. . . . " Larry D. Roberts, "A Preliminary Survey of Historic Southeastern Alaskan Fur Farming," (Working paper supplied by author, February 2010).
2. Fiorello LaGuardia, "Minority Report, Leasing Public Lands in Alaska for Fur Farming," Committee on the Public Lands, H.R. 678, 69th Congress, Session I, March 27, 1926.
3. Editorial, *Fairbanks Daily News-Miner*, October 2, 1926.
4. Chapter 745, 69th Congress, Session I, July 3, 1926.
5. George Parks to Ray L. Wilbur, 7 April 1932, Alaska State Archives, RG 101, SR 130, VS 334, folder 29-10. To approximate 2010 dollars according to Consumer Price Index, multiply 1926–1930 dollars by 12.7. www.measuringworth.com (accessed 4-9-11).
6. Mangusso and Haycox 1989, 14.
7. Like many new Alaska fur farmers, Parker was a WWI veteran. Before completing Kansas City Veterinary College, he spent a year as a private in Veterinary Company #1 caring for army mules and horses.
8. Susie Pederson, "Ed Ophiem, Sr., 'Fox Farming,'" *Elwani* 6 (1979): 9. Harvey 1991, 275 and 268.
9. "Basal C. Parker," obituary, *Journal of the American Veterinary Medical Association* 88 (January 1936), 557. Parker to H. W. Terhune, 30 November 1928, AK State Archives, Alaska Territorial Department of Audit, Fox Brand Program Files, RG 106, Series 102, Box 1, folder "General Correspondence 1923–39."
10. "Sourdough Notes," *The Pathfinder of Alaska*, December 1924, p. 16.
11. Ashbrook 1923, 6.
12. Ashbrook and Walker 1925, 31.
13. Foxes are naturally monogamous, and early fur farmers kept equal numbers of male and female breeders. Polygamous matings saved farmers money since they only needed to retain a few males to impregnate a large group of females.
14. Only about half the territorial governors were Alaskans at the time they were appointed. Parks was a long-term resident, leaving only for military service during WWI. He died in Juneau in 1984 at age 100.
15. Chapter 53, Alaska Session Laws of 1927.

16. Parks to Paul Redington, Chief of the Biological Survey, 17 May 1917. Redington to Parks (telegram and letter), 28 May 1927. AK State Archives, RG 101, SR 130, VS 274, folder 29-10a. "Territory Unable Engage Veterinary," *Fairbanks Daily News-Miner*, June 21, 1927.
17. Earl Graves, "Weekly Itinerary and Report of Activities, 1929," AK State Archives, RG 101, SR 726, VS 304, folder 29-10.
18. Graves, "Weekly Itinerary," 1928; and Graves to Guy Turnbow (telegram), April 23, 1929. Both documents in AK State Archives, RG 101, SR 726, VS 304, folder 29-10.
19. Earl Graves, "Alaska Fur Farming Conditions" (n.d.; text indicates it is his 1929 annual report), 1 and 8, AK State Archives, RG 101, SR 726, VS 289, folder 29-10a.
20. Fur prices in this chapter from Alaska Game Commission, "Furs Shipped from Alaska during the Year [year filled in]," annual press releases.
21. AGC, "Furs Shipped from Alaska during the Years 1925 and 1929." U.S. Bureau of Biological Survey, "Fur Farms of Alaska in Operation March 31, 1924," AK State Archives, RG 106, SR 102, VS 1, folder "Island Information." AGC, "Fur Farmers of Alaska Holding Licenses under the Alaska Game Laws for the year ending June 30, 1929," and AGC, "Annual Report to the Secretary of Agriculture, 1930," 25, AGC Collection, UAF Archives.
22. Snider 1929, 18–19.
23. Ashbrook 1927.
24. AGC, "Furs Shipped from Alaska during the Years 1924 and 1929."
25. J. Forester and A. Forester 1980, 38.
26. "Ship Foxes to Foreign Land," *Fairbanks Daily News-Miner*, October 28, 1924. "Many Silvers Being Shipped to Europeans," *The Fur Farmer Magazine* 3, no. 7 (April 1926): 39.
27. "Express Rates on Foxes" *The Fur Farmer Magazine* 3, no. 10 (December 1926: 23).
28. *Annual Report of the Governor of Alaska to the Secretary of the Interior, 1931*, p. 155.
29. In addition to blue, white, and silver foxes, Turnbow raised mink and chinchilla rabbits. He also tried to breed martens. His farm was a popular roadhouse stop.
30. "Those Who Come and Go," *Fairbanks Daily News-Miner*, November 13, 1930. "Fromm Bros., World's Largest Fur Ranchers, May Establish a Fox Farm in the Fairbanks District," *Fairbanks Daily News-Miner*, November 13, 1929.
31. "Morrison Loses Fox," *Nenana Daily News*, November 23, 1920. "Ships 240 Foxes to the Far East," *Fairbanks Daily News-Miner*, October 22, 1924. "Fox

Farming Getting Good," *Fairbanks Daily News-Miner*, August 4, 1924. "Fur Farming in Alaska," *The Fur Farmer Magazine* 1, no. 2 (March 1923): 11. J. Forester and A. Forester 1980, 122–123. Another silver fox franchise chain, Maeser Fur Farms based in Minnesota, had fewer farms but extended to "units" as far away as Finland, Bavaria, South Africa, and Tebenkof Bay in Alaska. "Domesticated Fur Farming Has Arrived," *The Fur Farmer Magazine* 5, no. 1 (December 1928): 14–15. Tebenkof Bay was Maeser's experiment in blue fox farming. It closed within a few years. "Experimental Fur Farm Has Been Started," *The Alaska Weekly*, October 7, 1927.

32. ACG, "Furs Shipped from Alaska during the Years 1925 through 1929." AGC, "Annual Report to the Secretary of Agriculture, 1930," 25, AGC Collection, UAF Archives. Earl F. Graves, "Territorial Veterinarian Report, February 1928," 8, AK State Archives, RG 101, SR 130, VS 274, folder 29-10a. Graves, "Alaska Fur Farming Conditions," (n.d., text indicates 1929 annual report), AK State Archives, RG 101, SR 726, VS 289, folder 29-10a.

33. Susie Pederson, "Ed Ophiem, Sr., 'Fox Farming,'" *Elwani*, 6 (1979): 9.

34. Hospitality extended to nicotine: "At one place . . . eight different tobaccos were brought out . . . boxes of cigars, cans of smoking tobacco, two kinds of chewing tobacco and some snuff." Graves, "Territorial Veterinarian Report for March 1928—13th–31st," 3, AK State Archives, RG 101, SR 130, VS 274, folder 29-10a.

35. Graves, "Report of Veterinarian Investigating Fox Ranching In Alaska, January 1928," 5–6, AK State Archives, RG 101, SR 130, VS 274, folder 29-10a. Parks to W. C. Henderson, 14 February 1930, AK State Archives, RG 101, SR 726, VS 304, folder 20-10.

36. In the 1920s, true chinchillas were not available to North American fur famers. Alaskans imported chinchilla rabbits, specially bred to produce a chinchilla-like coat. "Rabbits Form New Industry for Northland," *Petersburg Herald*, July 24, 1925. In the late 1930s, American-bred chinchillas became available, and some Alaskans raised them during the 1940s and 1950s.

37. In the national 1930 census, foreign-born whites (half of whom were from Scandinavia or Finland) comprised 17 percent of the population. U.S. Bureau of the Census, *Fifteenth Census of the United States, 1930, Outlying Territories and Possessions* (Washington, DC: GPO, 1932), 21.

38. Farmers in all four judicial districts were interviewed, but the challenges of transportation meant the sample was not random. U.S. Bureau of the Census, *Nonpopulation Census Schedules for Alaska, 1929: Agriculture*. Record Group 29, Microfilm M 1871, NARA.

Endnotes

39. U.S. General Land Office, Department of the Interior, "Register of Final Homestead Receipts and Fur Farm Rentals, July 1, 1899 to March 2, 1931," NARA. Other records as previously cited. Author's compilation.
40. Huston 1963, 70 and 72. "Fur Farming on Increase This Region," *Fairbanks Daily News-Miner*, December 9, 1929.
41. Graves, "Alaskan Fur Farming Conditions" (n.d., text indicates annual report 1929), pp. 1, 6, 8–9, AK State Archives, RG 101, SR 726, VS 289, folder 29-10a.
42. Graves, "Veterinarian's Report, June 1929," AK State Archives, RG 101, SR 726, VS 289, folder 29-10a.
43. Attorney General John Rustgard dismissed the Anchorage-based Western Fur Breeder's Association claim that Graves had to give fur farmers his undivided attention because the territorial veterinarian's legislated duties required it. Rustgard's opinion was that Graves had been hired for two part-time jobs—cattle tester and territorial veterinarian. "Fur Farmers Are Unable to Compel Governor to Act," *Anchorage Daily Times*, July 3, 1929.
44. Graves to Parks, 29 April 1930, AK State Archives, RG 101, SR 726, VS 289, folder 29-10a.

Chapter 6

1. AGC, "Furs Shipped from Alaska during the Years 1929, 1932, and 1933." Subsequent fur prices in this chapter from the same source. To approximate 2010 dollars according to Consumer Price Index, multiply 1930s dollars by 15.6. www.measuringworth.com (accessed 4-9-11).
2. Ashbrook 1930, 2, 10, 17, 33–43.
3. "Bulletin by Seattle Fur Exchange," *The Fur Farmer Magazine* 7, no. 4 (October 1931): 15.
4. "Alaska Commerce Declines," *Fairbanks Daily News-Miner*, January 16, 1933. Huston 1963, 73–74.
5. "Reviewing the Silver Fox Season," *The Fur Farmer Magazine* 7, no. 2 (October 1931): 15, 21, 22, 26.
6. Paul Redington, "Fur Farming," *The Fur Farmer Magazine* 7, no. 7 (January 1932): 4.
7. "The Editor's Pen," *The Fur Farmer Magazine* 7, no. 9 (March 1932): 24.
8. AGC, "Annual Report to the Secretary of Agriculture, 1935," p. 28, Box 2, folder 8, AGC Collection, UAF Archives. "Annual Report to the Secretary of Agriculture, 1929," p. 22, Box 1, folder 3, AGC Collection, UAF Archives.

Endnotes

AGC, "Fur Farmers of Alaska Holding Licenses under the Alaska Game Laws for the Year Ending 1929."

9. Soon after *The Fur Farmer Magazine* ended publication, Charles Garfield established *The Fur Journal*, describing it as the "Official Organ of the Puget Sound Fur Farmers Association and Associated M&M Fur Farmers, Inc." (The latter comprised Milligan and Morrison franchise farmers and the Rocky Mountain Fur Growers Association.) *The Fur Journal* carried Alaska news, but its focus and advertising were nationwide. It ended publication in 1953; Garfield was its sole editor.

10. "Inspection Report of Auditor Wilt, Period 8/1/33–7/13/35" and "Report of Office Inspection by W. C. Ellis 2/25/34–3/18/34," U.S. Forest Service, Records 1870–2000 (Inspection Reports 1909–1963), NARA. The 50 percent reduction in fees began in 1932.

11. "Memo for Mr. Loving on Wilts Report 9/13/35," unsigned, p. 3, U.S. Forest Service, Records 1870–2000 (Inspection Reports 1909–1963), NARA. Alice Clock was able to continue living on her island until her death in 1973. Janson ca. 1985, chapter 8, p. 3.

12. Bunnell to Parks, 11 June 1930, AK State Archives, RG 101, SR 276, VS 289, folder 29-10a. Loftus's ties to Alaska included two brothers in Fairbanks. Near the Fairbanks campus he also owned a 172-acre homestead, which would eventually be purchased by Charles Bunnell.

13. Parks to Loftus, 12 June 1930, Alaska State Archives, RG 101, SR 276, VS 289, folder 29-10a.

14. Loftus ca. 1970.

15. Jule B. Loftus, "Report of the Territorial Veterinarian, September 1932," AK State Archives, RG 101, SR 726, VS 350, folder 29-10.

16. Loftus to Parks, 18 April 1931, AK State Archives, RG 101, SR 130, VS 319, folder 29-10.

17. Loftus ca. 1970.

18. Loftus, "Report of the Territorial Veterinarian" (undated but identified in text as annual report, 1931), 6, AK State Archives, RG 101, SR 130, VS 334, folder 29-10.

19. Viscera and other waste parts from game were permissible food for captive furbearers. Predatory birds specifically listed were eagles, ravens, crows, owls, and cormorants.

20. Lois Hudson Allen, "Trail Lake Fur Farm," *The Alaska Sportsman* 8, no. 3 (March 1942): 10.

21. Loftus, "Report of the Territorial Veterinarian, April 1931," AK State Archives, RG 101, SR 130, VS 319, folder 29-10. Territorial veterinarians advised farmers whose foxes developed rickets to add cod-liver oil to the feed.
22. This waste included not only heads, fins, and tails but also halibut cheeks and "short" cans of salmon that weighed less than the packer's label indicated.
23. In 1939, the seal bounty was increased to $3. The program finally ended in southeast and southcentral Alaska in 1967 and was terminated in northwest Alaska in 1971.
24. A dead whale delivered to the beach was cut into small pieces, lightly salted, and stored in tanks or underground pits. "From the Blue Fox," *The Fur Farmer Magazine* 3, no. 7 (April 1926): 14. Huston 1963, 61. Frank Cunningham, a farmer near Seward, installed derricks and huge cooking vats to make use of whale meat. "Fox Industry Is Profitable," *The Alaska Weekly*, August 21, 1925.
25. *Annual Report of the Governor of Alaska to the Secretary of the Interior*, years 1913–1933 (Washington DC: GPO).
26. Complaints that Bowman was taking advantage of the Aleuts were sent to government officials by both Ernest P. Stowell, an Atka school teacher, and fox farmer Durell Finch, who was also U.S. Commissioner at Unalaska. Bowman was said to have bought two village stores "for a song," then jacked up prices on food and boats he sold to two communities. The result was that everyone was in debt to him. The *Umnak Native*, which he sold to the Umnak village co-op, had been modified by replacing a newer engine with an older one. A two-year government investigation substantiated the complaints, but cleared him of definable criminal activity. Finch to H. W. Terhune, 18 October 1930, reel 271, Territorial Governors Correspondence. Morgan 1992, 126.
27. *The Fur Farmer Magazine* 7, no. 2 (August 1931): cover photo and caption.
28. Several years earlier, Bowman had claimed that his company was the largest blue fox farm in the world, according to *The Fox Breeders Gazette* 2, (1925): 40.
29. Alaska Commercial Company Records 1888–1940, Stanford University, JL006, Series 1, Subseries B, Box 14, Folder 2, Account 607 (Kanaga Ranching Company mortgage 1932) and Account 612 (Umnak Natives and Kanaga Ranching Company agreement, 1926). Kanaga Ranching owned a number of boats used for fur farming and to supply their stores on Kanaga and Atka Islands.
30. Janson ca. 1985, chapter 3, pp. 4 and 11–12.
31. Jahn 1947, 18. Although American Karakuls were cross-bred with domestic sheep, their ancestors were adapted to cold arid climates such as Afganistan, and they did not thrive in the wet coastal weather of the Aleutians. Market prices were rarely high enough to cover the cost of caring for the animals and shipping the pelts.

32. "Fur Farming to Save Matanuska Project," *The Fur Trade Journal* 2, no. 6 (April 1936): 1–2. "First Colonist to Leave 1939," *Fairbanks Daily News-Miner*, January 17, 1939.

33. Ohmer's partners in the farm were Fred Porter, Charlie Craig, and Jess E. Ames.

34. "Fur Farmers Are Circularized," *Petersburg Press*, November 6, 1936. Ohmer came to the territory from Montana in 1915, started farming foxes in 1920, and then switched to mink.

35. "Goshaw Talks on Fur Farms," *Nome Nugget*, March 22, 1930. "Pelts from the Farmer: Fur Experiment Farm Needed," *The Fur Farmer Magazine* 5, no. 6 (May 1929): 27. "Alaska Fur Experimental Station," *The Fur Journal* 3, no. 2 (December 1936): 10.

36. Chapter 42, Alaska Session Laws of 1937. Victor Ross, Loftus's pilot on the trip to the Noatak River and Shishmaref, was a representative in the territorial legislature that passed the bill.

37. "Alaska's Experimental Fur Farm," editorial, *Fairbanks Daily News-Miner*, October 15, 1938. Arndt 1978, 3. Loftus to Loren Oldroyd, 5 May 1938, Alaska Agricultural Experiment Stations Collection, Box 6, folder 11, UAF Archives.

38. The 1932–1933 nadirs were approximately $21 for blue fox, $6 for mink, and $41 for silver fox. By 1938, Alaska blue fox pelts sold at approximately $26.50, mink at $11.50. The 1939 silver fox price was $26.50.

39. "Pelt Crop on Silver Fox Farms 28 Times Larger than in 1923," *The Fur Journal* 2, no. 4 (February 1936): 7. "Trade Menace of Foreign Silver Fox," *The Fur Journal* 3, no. 6 (April 1937): 3. "Fur Farming Developments in Germany," *The Fur Journal* 4, no. 8 (June 1938): 6–7. In 1936, European countries sold 16,000 pelts in the United States.

40. AGC, "Minutes of the Annual Meeting, 1933, 1934, and 1942," MF 57 Alaska State Library. Raccoons also escaped captivity and survived for a decade or more on Japonski Island near Sitka and on Long Island near Kodiak. A few descendants still remain in Alaska. A 2005 statement from biologists reads: "The current population of raccoons is very small and not considered a threat to coastal Alaskan ecosystems." Barbara Schrader and Paul Hennon, *Assessment of Invasive Species in Alaska and Its National Forests*, www.akweeds.uaa.alaska.edu/pdfs/literature/Schrader_R10_Inv_Spp_Assessment.pdf (accessed 5-2-10).

41. Louis G. Scott to Mr. Scheffer, U.S. Fish and Wildlife Service, 13 February 1947, Series 7, Box 11, Folder 8, Ashbrook Papers, Smithsonian Archives.

42. AGC, "Fur Farmers of Alaska Holding Licenses, 1941."

43. Harvey 1991, 278. Probate Court for the Kodiak Precinct, Third Division, Territory of Alaska, Case #142, In the Matter of the Estate of Basal C. Parker, Final Report June 20, 1936.
44. J. Forester and A. Forester 1980, 122–123. M&M's competitors, the Maeser chain, also survived the Depression, although the Tebenkof Bay farm closed. C. G. Maeser to Juneau Forest Service, 4 August 1942, U.S. Forest Service Records, Land Case Files 1907–1989, NARA.
45. Hegness died in Washington state in 1963. Freuchen 1953, 242. "Winner of First Nome Big Dog Race Now with M.-G.-M," *Fairbanks Daily News-Miner*, June 28, 1933.
46. U.S. Forest Service, Land Case Files 1907–1989, NARA.
47. "Fur Farm Inspection Report, C. E. Zimmerman, March 12, 1941," Box 39, Forest Service Land Case Files, 1920–1974, NARA.
48. "Big Fox Farm Goes to Kenai Lake Country," *The Alaska Weekly*, September 6, 1929.
49. Sale contract dated April 27, 1929, Box 2, Wilson Fiske Erskine Collection 1884–1964, UAF Archives. W. J. Erskine retired from his other Kodiak businesses to move to San Francisco in the mid-1940s. He died there in December 1948.
50. D. Tewksbury and W. Tewkesbury 1947, 188. Larry D. Roberts, "A Preliminary Survey of Historic Southeastern Alaskan Fur Farming," manuscript February 2010, Background file XMF: Lemesurier Island, Ibach & Co.
51. Janet Leekley Eddy, personal communication with author, January 2010, Juneau.
52. The new logo came into use about 1940. It was often accompanied by the Seattle Fur Exchange motto, "An organization devoted to safeguarding fur farmers' interests." Advertisement, *American Fur Breeder* (January 1944).
53. Frank Dusfrense quoted in "White Foxes Being Raised in the Arctic," *Petersburg Herald*, July 24, 1925. Mae Brandon, telephone conversation with the author, August 2009.
54. AGC, "Fur Farmers of Alaska Holding Licenses, 1941."
55. "U.S. Concludes New Quota Agreement," *American National Fur & Market Journal*, January 1941, 5.

Chapter 7

1. "Urata Returns Home From Trip to Japan," *The Wrangell Sentinel*, May 26, 1939; "Local Items," *The Wrangell Sentinel*, June 4, 1937. Telephone conversations with Masuye Urata, daughter-in-law of Ryotaro Urata, January and February 2008.

To approximate 2010 dollars according to Consumer Price Index, multiply 1941–1945 dollars by 13. www.measuringworth.com (accessed 4-9-11).

2. Murie 1959, 300–302. Natives told Murie that foxes also took spawning salmon by leaping into the water.

3. Turner 1886, 303–304. Much later, a few Aleutian geese would be discovered breeding on two other islands.

4. Edgar P. Bailey, "Alaska's Alien Animals," in *Pacific Seabird Group Bulletin* 20, no. 2 (1993): 5–8.

5. "West Alaska Made into Extensive Reserve," *Fairbanks Daily News-Miner*, November 2, 1939.

6. Jack Martin, Executive Secretary Aleutian Islands Fur Farmers Association, to Governor Ernest Gruening, 4 September 1940, reel 271, Territorial Governors Correspondence.

7. Ira Gabrielson, Director U.S. Fish and Wildlife Service, to Anthony Dimond, 13 October 1940, reel 271, Territorial Governors Correspondence.

8. Among the fifteen board members were Harold Bowman (Kanaga Ranching); Durell Finch (U.S. Commissioner in Unalaska, whose fur farm career began in Kodiak with W.J. Erskine); Clara Goss (widow of A.C. Goss, holder of licenses for five islands); and Capt. C.T. Pedersen (Aleutian Fur Co.). Although two Aleuts, Nicholai Bolshanin and Sergias Golley, were also on the board, attorney Martin's letter generally disparaged Native fur farmers and made much of the white man's burden. He attributed the success of the Irish-Aleut brothers, William and Hugh McGlashan of Akutan, to "the preponderance of white blood in them [which] far outshines the lassitude common among the native bloods." Martin to Gruening, 4 September 1940, reel 271, Territorial Governors Correspondence.

9. Martin to Gruening, 4 September 1940, reel 271, Territorial Governors Correspondence.

10. Frank Dufresne, Executive Officer of the Alaska Game Commission, wrote, "The facts gathered by me strongly indicated that the native Aleuts were being exploited by the white traders, several of whose names now appear on the executive board of the Aleutian Islands Fur Farmers Association." Dufresne to Gruening, 29 October 1940. Similar sentiments were expressed by Charles W. Hawkesworth, Office of Indian Affairs, to Gruening, 23 October 1940; and Gabrielson to Dimond, 13 October 1940. All letters from reel 271, Territorial Governors Correspondence.

11. Christopher Cueva, "America's Territory: The Aleut Evacuation—A Grave Injustice," essay posted at www.akhistorycourse.org (accessed 4-13-10).

12. Before the Japanese left the Aleutians, Governor Gruening sent "a plea" to Ira Gabrielson of the Biological Survey to send the Pribilof Aleuts home "at the

earliest possible moment. I am convinced that a winter in the Pribilofs would be infinitely preferable to another winter at Funter Bay." The governor had visited Funter Bay after receiving a letter from his attorney general, Henry Roden, who wrote, "I have no language at my command which can adequately describe what I saw [at Funter Bay]. . . . I cannot understand how public authorities can tolerate conditions existing there." Roden to Gruening, 23 September 1943; Gruening to Roden, 22 September 1943; Gruening to Gabrielson, 22 September 1943. All letters from reel 271, Territorial Governors Correspondence. Aleuts from islands other than the Pribilofs had to wait to return home until after Japan surrendered.

13. Pribilof Aleuts would finally be granted U.S. citizenship in 1966. The 1971 Alaska Native Claims Settlement Act gave them control of their land. In 1978, they received $8.6 million in partial compensation for their "unfair and unjust" treatment by the federal government between 1870 and 1946. In 1988, all Aleuts received a government apology and restitution funds as a result of their WWII internment in southeast Alaska.

14. "Washington Committee Designs Fur Lined Aviator's Uniform," *American National Fur & Market Journal* (July 1942): 7–8. "Hari-Kari Is Not Victory," *The Fur Journal* 8, no. 11 (September 1942): 11.

15. "Hari-Kari Is Not Victory," *The Fur Journal* 8, no. 11 (September 1942): 11.

16. Ralph Robey, "Fur Coats and Winning the War," *The Fur Journal* 9, no. 1 (November 1942): 15.

17. In 1942, this tax brought the government approximately $33 million; in 1945, it provided $86 million. Fur farmers and manufacturers pointed to it as evidence of the value of the industry to the country. "Internal Revenue Collected from Retailer's Excise Tax on Fur Apparel," Series 3, Box 3, folder 13, Ashbrook Papers, Smithsonian.

18. This price is an average for all grades of silver pelts from Alaska. Fur prices in this chapter are from AGC, "Furs Shipped from Alaska During the Year. . .," annual press releases. Nearly all of the blue and silver fox and the majority of the mink were farmed rather than trapped.

19. In 1915, the U.S. horse population was 21.4 million; in 1945, it was 8.3 million. "Horse Population Declines," *American Fur Breeder* (June 1946): 20.

20. *Alaska Agricultural Experiment Stations: 11th Progress Report, 1946* (Fairbanks: University of Alaska, 1946), 44–54. Lorin Oldroyd to Governor Gruening, report, "Work Summary: Alaska Experiment Stations, August 16, 1944," pp. 7–8, Alaska Agricultural Experiment Station Collection, UAF Archives.

21. "From Ketchikan to Barrow," *The Alaska Sportsman* 9, no. 4 (April 1943): 18. James R. Leekley, "Petersburg Station Seeks to Establish Standards for Fur Farming Industry," *Fairbanks Daily News-Miner*, September 9, 1943.

22. AGC, "Minutes of the Annual Meeting of the Alaska Game Commission," February 1945, pp. 5, 12.

Chapter 8

1. Gorman 1947, 11.
2. George L. DeVenne, "Blue Gold of the Aleutians," *The Alaska Sportsman* 5, no. 7 (July 1939): 8–9, 22–24.
3. Christian T. Pedersen Papers 1920–1966, Consortium Library Archives, Anchorage.
4. V. H. Wilson, quoted in Huston 1963, 98.
5. T. Paul, 2009, 101–102 and 121–122. Recent studies suggest that the return of guano to these islands has encouraged the regrowth of the indigenous plants that declined in tandem with the bird population. "This Week in Science," *Science* 307, no. 5717 (March 25, 2005): 1835.
6. Schlung and Nikolski students 2003, 38–39. Swanson and high school students at Unalaska 2003, 68–69. To approximate 2010 value according to Consumer Price Index, multiply 1946–1959 dollars by 7.3, and 1960–1980 dollars by 5.5 www.measuringworth.com (accessed 4-9-11).
7. Series 6, Box 7, folder 3, Ashbrook Papers, Smithsonian.
8. Seton H. Thompson, *Alaska Fishery and Fur-Seal Industries, 1947, 1948, 1949* (Washington, DC: GPO).
9. Ashbrook to Chief of Wildlife Research, memorandum, 5 December 1950, Series 7, Box 8, folder 5, Ashbrook Papers, Smithsonian.
10. AGC "Minutes of the Annual Meeting, February 11, 1941," MFilm 57, Alaska State Library. After WWII, Goshaw sold 75 of his "platinum" fox pups as breeders to Fromm Bros. of Wisconsin. Denison 1950, 151. Goshaw had also attempted to cross *Vulpes* silver foxes with *Alopex* white foxes without success. Later, specialists were able to accomplish this with artificial insemination, but the offspring were sterile.
11. Ashbrook to Director, Fish & Wildlife Service, memorandum, 18 April 1947; James Leekley to Director, Fish & Wildlife Service, 19 July 1946; Seton H. Thompson to Clarence L. Olson, 19 October 1953. Series 7, Box 8, folders 5, 7, 8, Ashbrook Papers, Smithsonian.
12. Seton H. Thompson, *Alaska Fisheries and Fur Seal Industries, 1950, 1951, 1952, 1952, 1954, 1955* (Washington, DC: GPO).
13. Olsen to Chief Alaska Fisheries, memorandum, 23 February 1954, Series 7, Box 8, folder 5, Ashbrook Papers, Smithsonian.

Endnotes

14. Report republished as Goldschmidt and Haas 1998, 34, 47, 56, 58, 77, 86–87, 92, 94, 110, 113, 122, 127, 132–136, 164, 172–173, 176–177.
15. E. E. Carter, "Inspection Report, September 20, 1923," U.S. Forest Service Records, 1870–2000, Box 1, NARA.
16. George Dalton of Hoonah relates that in 1921 he and his father went to an island in Icy Strait to plant potatoes in a garden that they had long cultivated and fertilized with seaweed. They were chased away by a fox farmer with a gun who told them, "Get the hell off my island. . . . I got permits from the Forest Service and the Land Office to be here." F. Paul 2003, 81.
17. U.S. Forest Service Land Case Files 1907–1989, NARA.
18. Oliver T. Edwards, "Report on Fur Farm Inspection," November 24, 1939. Oliver T. Edwards Papers, Archives and Special Collections, University of Alaska Anchorage.
19. Heintzelman to Jacobsen, District Ranger, Cordova, 22 August 1940, reel 280, Territorial Governors Correspondence.
20. Huston 1963, 89.
21. Bailey 1993, 30–31. This source also notes that near the Aleutian island of Sanak, some small, unnamed, abandoned islands were swept clean of foxes by a 1946 tsunami that reached 30 meters at Unimak Island.
22. Blair 1923.
23. Frank G. Ashbrook, "Trade Names in the Fur Industry," *Journal of Mammalogy* 4, no. 4 (November 1923): 218–219.
24. Over a million chinchilla pelts were believed sold in Europe from 1899 to 1901. The American engineer was M. F. Chapman. See "The Mathias F. Chapman Story" at www.edchinchillas.co.uk/The_Chin/chapman.html (accessed 7-7-11). Janson ca. 1985, chapter 16, p.5.
25. Trade names of the Mutation Mink Breeders Association (EMBA) also included Desert Gold (light brown), Argenta (grey), Cerulean (blue), Jasmine (white), Tourmaline (pale beige), and Aeolian (grey taupe). "Fur Farming Special Feature No. 1: True Colors" at www.furcommission.com (accessed 4-27-10).
26. Samet 1950, 118.
27. This is a slight exaggeration. According to the USDA, during 1994, mink prices bumped up from $33 to $53 but promptly fell the next year back to $35. "Mink Production in the United States, 1974–2008," www.furcommission.com (accessed 5-8-10). During the current century, prices have been in the $50 to $65 range.
28. Fuchs 1957, 75–81, 139.
29. "New Law to Ban Soviet Fur Imports Into United States," *Fairbanks Daily News-Miner*, June 4, 1951. Fuchs 1957, 69–70.

30. In 1945, Pan American World Airlines charged $81.40 for hauling 167 muskrat pelts worth $835 from Fairbanks to Seattle. Huston 1963, 112.
31. Huston 1963, 87–89.
32. "Industry of the Past," *Raincountry* series, videotape 911, KTOO TV, 1994. Bob Kederick, "All Around Alaska: Fur Farms Losing Out," *Anchorage Daily Times*, November 15, 1957.
33. "Oldroyd Says That Small Fur Farms Way to Security: Domestic Fur and Garden Farming Can Be Combined to Provide Living," *Seward Gateway*, February 27, 1940.
34. Telephone conversation on May 25, 1911, with Henry Ivanoff, one of the two islanders in charge of the Nunivak mink farm.
35. "The Fur Industry in Alaska," *Review of Business and Economic Conditions* 3, no. 4 (November 1966): 6–7.
36. "Link's Minks Vanquished by Presidential Veto," *Fairbanks Daily News-Miner*, April 25, 1949. "Mink Farmers to Get Day in Court," *Fairbanks Daily News-Miner*, August 3, 1949. No case by the three plaintiffs—Hilda Links, E. J. Ohman, and Fred L. Kroesing—appears in legal indices; nor was a case file found in territorial district court records for 1947–1954 stored at NARA.
37. D. E. Stubbs, "About Worms That Infest Fur Bearers," *Seward Daily Gateway*, October 23, 1931. Stubbs had an "inholding" in what is now called Denali National Park because the park boundaries expanded to surround his land in 1932.
38. "Alaska Fox Farmer Awarded Damages for Loss of Land and Animals," *The Fur Journal* 4, no. 9 (December 1937): 14. "D. E. Stubbs Passes Away in the States," *Fairbanks Daily News-Miner*, January 31, 1939.
39. "Study Set of Mink's Reaction to Booms," *Fairbanks Daily News-Miner*, October 15, 1969.
40. Travis and Leekley 1972.
41. Series 4, Box 4, folder 4, Ashbrook Papers, Smithsonian.
42. "Eighth Annual Humane Trap Contest," *The Fur Farmer Magazine* 9, no. 12 (December 1934): 2. In 1959, contest prizes totaled $20,000.
43. J. E. Shillinger, "Humane Aspects in Fox Farming," speech, American Society for Prevention of Cruelty to Animals, St. Louis, MO, September 30, 1929. "Fromm Bros., World's Largest Fur Ranchers," *Fairbanks Daily News-Miner*, November 13, 1929.
44. Spector 1998, 55–62, 80.
45. Underwood and Baldrige 1981. Eddy 1983. *A Village Fur Farm* (Fairbanks: Interior Village Association Alternative Development Program, 1982).

46. "A Look at Alaskan Fur Farming," *Alaska Farm and Garden* 2, no. 8 (September 1982): 14–16. Strohmeyer 1993, 113. *Jim Rice v. Mike Johnson*, U.S. Bankruptcy Court for the District of Alaska, Case No. A92-00602-DMD, February 4, 1994.
47. Spector 1998, 83.
48. Joni Scharfenberg, "Fur Farming in Alaska," *Alaska Farm Magazine* 2, no. 1 (January 1982): 36–37. Telephone conversations with Irene Christie and Vickie Greenleaf, May 1, 2010. Whitestone may not be the last fur farm in Alaska. In 2010 mink prices were approximately $80 per pelt, and at least one pair of enterprising Alaskans, who wished to remain anonymous, was looking for land, potential food supplies, and customers for the mink they planned to breed. Personal communication, June 2010.

REFERENCE LIST

Alaska Agricultural Experiment Stations: Progress Report, various years. Fairbanks: University of Alaska, various dates.

Alaska Commercial Company Records 1888–1940. Account 607 (Kanaga Ranching Company mortgage 1932) and Account 612 (Umnak Natives and Kanaga Ranching Company agreement 1926). Stanford University, JL006, Series 1, Subseries B, Box 14, Folder 2.

Alaska Game Commission. "Annual Report to the Secretary of Agriculture," various years, 1925–1948. Alaska Game Commission Collection, University of Alaska Fairbanks Archives.

———. "Fur Farmers of Alaska Holding Licenses under the Alaska Game Laws for the Year Ending . . . ." Annual license lists, various years 1925–1951.

———. "Furs Shipped from Alaska During the Year . . . ." Annual press releases, 1925–1951.

———. "Minutes of the Annual Meeting," various years 1925–1948. MFilm 57, Alaska State Historical Library, Juneau.

———. "Report of the Executive Officer to the Alaska Game Commission for the Period . . ." MFilm 56, Alaska State Historical Library, Juneau.

Alaska Territorial Governors. General Correspondence of the Alaskan Territorial Governor, 1909–1958. MFAR 27, Alaska State Historical Library, Juneau.

Allen, Lois Hudson. "Trail Lake Fur Farm." *The Alaska Sportsman* 8, no. 3 (March 1942): 10–11, 23–24.

Alton, Tom L. "File: W. J. Erskine: Early Alaska Merchantilist," manuscript, n.d. Biography File: W. J. Erskine, Alaska State Historical Library, Juneau.

Annual Report of the Chief of the Bureau of Biological Survey, 1921. Washington DC: GPO, 1921.

Annual Report of the Commissioner of Fisheries to the Secretary of Commerce for the Fiscal Year. Washington, DC: GPO, various years FY 1915–1919.

Annual Report of the Governor of Alaska to the Secretary of the Interior. Washington, DC: GPO, various years, 1913–1952.

Annual Report of the Secretary of the Interior for FY 1922. Washington, DC: GPO, 1922.

Applegate, Samuel. Samuel Applegate Papers, 1892–1925. Alaska State Historical Library, MS 3.

Arndt, Katherine, 1978. *The Alaska Experimental Fur Farm: Determination of Eligibility*. U.S. Forest Service, Stikine Area.

Ashbrook Frank G. F.G. Ashbrook Papers, circa 1915–1965. Record Unit 7143, Smithsonian Institution Archives, Washington, DC.

Ashbrook, Frank G. 1930. *Fur Resources of the United States: A Special Report to Supplement the Exhibit of the United States Government at the International Fur-Trade Exposition, Leipzig Germany, 1930*. Washington, DC: GPO.

———. 1927. *Mink Raising*. Washington, DC: GPO.

———. 1923. *Silver-Fox Farming*, USDA Bulletin No. 1151. Washington, DC: U.S. Department of Agriculture.

———. "Trade Names in the Fur Industry." *Journal of Mammalogy* 4, no. 4 (November 1923): 216–220.

Ashbrook, Frank G., and Ernest Walker. 1925. *Blue Fox Farming in Alaska*, USDA Bulletin 1350. Washington, DC: U.S. Department of Agriculture.

Bailey, Edgar P. 1993. *Introduction of Foxes to Alaskan Islands—History, Effects on Avifauna, and Eradication*, U.S. Fish and Wildlife Service Resource Publication 193. Washington, DC: GPO.

Bancroft, Hubert Howe. 1960. *History of Alaska 1730–1885*. 1886. Reprinted with Introduction by Ernest Gruening. New York: Antiquarian Press Ltd.

Barker, Kay. "Voluntary Exile." *The Alaska Sportsman* 5, no. 11 (November 1939): 8–9, 30–33.

Barrow, Mark V. Jr. 1998. *A Passion for Birds: American Ornithology after Audubon*. Princeton, NJ: Princeton University Press.

Beach, Rex. 1940. *Personal Exposures*. New York: Harper and Bros.

———. 1939. *Valley of Thunder*. New York: Farrar and Rinehart.

Berkh, Vasilii Nikolaevich. 1974. *A Chronological History of the Discovery of the Aleutian Islands*, Translated by Dmitri Krenov. Edited by Richard A. Pierce. Kingston, Ontario: The Limestone Press.

Berry, F. "Fur—Our Great Future Industry." *The Pathfinder of Alaska* 4, no. 8 (June 1923): 1–3.

Black, Lydia T. 1984. *Atka: An Ethnohistory of the Western Aleutians*. Kingston, Ontario: The Limestone Press.

———. 1999. *The History and Ethnohistory of the Aleutians East Borough*. Fairbanks, AK: Limestone Press.

———. 2004. *Russians in Alaska, 1732–1867*. Fairbanks: University of Alaska Press.

Blair, George H. 1923. *Illustrated Catalogue of Fur Bearing Animals*. Boston: Jacob Norton's Sons Co.

Bower, Ward T. *Alaska Fisheries and Fur Industries in....* Washington, DC: GPO, various years 1918–1938.

Bower, Ward T., and Henry D. Aller. *Alaska Fisheries and Fur Industries in....* Washington, DC: GPO, various years 1915–1918.

Buskirk, Stephen W., and Philip S. Gipson. 1980. "Zoogeography of Arctic Foxes (*Alopex lagopus*) on the Aleutian Islands." In *Worldwide Furbearer Conference Proceedings*, Vol. I. Edited by Joseph A. Chapman and Duane Pursley. Frostburg, MD: Worldwide Furbearer Conference.

Cadzow, Dan [misspelled "Don"]. "Fur Farming above the Arctic Circle," *Fur-Fish-Game*, March 1917, 16–17.

Cameron, Jenks. 1929. *The Bureau of Biological Survey: Its History, Activities and Organization*. Reprint, Baltimore: Johns Hopkins Press, 1974.

Chevigny, Hector. 1965. *Russian America: The Great Alaskan Venture 1741–1867*. New York: Viking Press.

Coxe, William. 1996. *Account of the Russian Discoveries between Asia and America*. New York: Argonaut Press Ltd., originally published 1787.

Crompton, C. E. 1920. "Methods Employed in the Management of Blue Foxes." Report to the Commissioner of Fisheries, October 23, 1920. U.S. Fish and Wildlife Collection, MS 51, Box 2, Folder 5. Alaska State Historical Library, Juneau.

Crosby, Albert W. 2003. *America's Forgotten Pandemic: The Influenza of 1918*, 2nd ed. Cambridge: Cambridge University Press.

Cueva, Christopher. "America's Territory: The Aleut Evacuation—A Grave Injustice," essay posted at www.akhistorycourse.org.

Dall, William H. 1870. *Alaska and Its Resources*. Boston: Lee and Shepard.

Dearborn, Ned. 1916. "Fur Farming as a Side Line." In *Yearbook U.S. Department of Agriculture, 1916*, 489–509. Washington, DC: GPO.

———. "Report on Fur Farming in Alaska, Summer 1915." F. G. Ashbrook Papers, circa 1915–1965, Record Unit 7143, Series 10, Box 15, Folder 8, Smithsonian Institution Archives.

Denison, B. W. 1950. *Alaska Today*. Caldwell, ID: The Caxton Printers, Ltd.

DeVenne, George L. "Blue Gold of the Aleutians." *The Alaska Sportsman* 5, no. 7 (July 1939): 8–9, 22–24.

Dmytryshyn, Basil, E. A. P. Crownhart-Vaughan, and Thomas Vaughan, editors and translators. 1988. *Russian Penetration of the North Pacific Ocean 1700–1797: A Documentary Record*, Vol. 2. Portland: Oregon Historical Society Press.

Eby, S. C. "A Fox Farm on an Island of the Sea." *Alaska-Yukon Magazine* 12, no. 1 (Feb. 1912): 7–10.

Eddy, Janet Leekley. 1983. "Fur Farm Production and Potential in Alaska." Report for Division of Finance and Economics, Alaska Department of Commerce and Economics.

Edwards, Oliver T. 1939. "Report on Fur Farm Inspection." Oliver T. Edwards Papers, Archives and Special Collections, University of Alaska, Anchorage.

Elliott, Henry W. 1976. *The Seal-Islands of Alaska*. 1881. Reprint, Kingston, Ontario: Limestone Press.

Erskine, W. J. "Report of the Kodiak Fox Farm, December 1, 1915," bound typescript. F. G. Ashbrook Papers, circa 1915–1965. Smithsonian Archives, Record Unit 7143, Series 7, Box 9, Folder 6.

Evermann, Barton Warren. 1914. *Alaska Fisheries and Fur Industries in 1913: Appendix II to the Report of the U.S. Commissioner of Fisheries for 1913*. Washington, DC: GPO.

Experimental Fur Farm of the Biological Survey, U.S. Department of Agriculture, Leaflet No. 6. Washington DC: GPO, July 1927.

Fisher, Robin, and Hugh Johnson, editors. 1993. *From Maps to Metaphors: The Pacific World of George Vancouver*. Vancouver: UBC Press.

Forester, Joseph E., and Anne D. Forester. 1980. *Silver Fox Odyssey: History of the Canadian Silver Fox Industry*. Charlottetown, P.E.I., Canada: Irwin Printing.

Freuchen, Peter. 1953. *Vagrant Viking*. New York: J. Messner.

Fuchs, Victor R. 1957. *The Economics of the Fur Industry*. New York: Columbia University Press.

"The Fur Industry in Alaska." *Review of Business and Economic Conditions* 3, no. 4 (November 1966).

Goldschmidt, Walter R., and Theodore H. Haas 1998. *Haa Aani, Our Land: Tlingit and Haida Land Rights and Use*. Editor Thomas F. Thornton. Seattle: University of Washington Press. Originally issued as government document "Possessory Rights of the Natives of Southeastern Alaska," 1946.

Gorman, Richard F., ed. 1947. *Fur Farming Opportunities in Alaska*. Juneau: Alaska Development Board.

Greely, A.W. 1925. *Handbook of Alaska: Its Resources, Products, and Attractions in 1924*. 3rd ed. Port Washington, NY: Kennikat Press.

Grosvold, Andrew. "Blue Fox Ranching on the Westward Islands." *The Pathfinder* 3, no. 7 (June 1922): 5.

Harding, A.R. 1909. *Fur Farming: A Book of Information on Raising Furbearing Animals for Profit*. Columbus, OH: A.R. Harding, Publisher.

Harvey, Lola. 1991. *Derevnia's Daughters: Saga of an Alaskan Village*. Manhattan, KS: Sunflower University Press.

Heideman, Charles W. H. 1909. *A Monograph of the Silver Fox*. Reprinted in "Taking the Pelt—The Earth's Oldest Industry, Part 15." *The Fur Farmer Magazine* 5, no. 7 (June–July 1929): 14–15, 31.

———. 1910. *The Story of the Silver Fox*. Seattle: The Ivy Press.

Henshaw, Henry W. 1912. "Report of the Chief of the Bureau of Biological Survey," *Alaska Agriculture* vol. 3, *Reports and Documents 1908–1912*. Washington DC: U.S. Department of Agriculture.

House Committee on Expenditures in the Department of Commerce and Labor, *Hearings on H.R. Res. 73 to Investigate the Fur Seal Industry of Alaska* (Washington DC: GPO, 1911) 340–342.

Huston, John Robert. 1963. "A Geographical Analysis of the Fur Farming Industry in Alaska." Master's thesis, University of California.

"Industry of the Past," *Raincountry*, videotape 911, KTOO TV, 1994.

Jahn, Karl. 1937. *Karakul Fur Sheep Breeding*. Chicago: Breeder Publications.

Janson, Lone. ca. 1985. "Those Alaska Blues: A Fox Tale," bound typescript.

Johnston, Samuel P. ca. 1940. *Alaska Commercial Company 1868–1940*. San Francisco: Edwin E. Wachter.

Jones, Cherry Lyons. 2006. *More than Petticoats: Remarkable Alaska Women*. Guilford, CT: Globe Pequot Press.

Jones, Dorothy Knee. 1980. *A Century of Servitude: Pribilof Aleuts Under U.S. Rule*. Washington, D.C.: University Press of America.

Jones, E. Lester. 1915. *Report of Alaska Investigations in 1914*. Washington, DC: GPO.

Judge, James. 1913. "The Blue Foxes of the Pribilof Islands." In J. Walter Jones, *Fur Farming in Canada, 60–69*. Montreal: Gazette Printing Co. Ltd.

Kederick, Bob. "All Around Alaska: Fur Farms Losing Out." *Anchorage Daily Times*, November 15, 1957.

Khlebnikov, Kiril Timofeevich. 1994. *Notes on Russian America, Parts II–V: Kad'iak, Unalashka, Atkha, the Pribylovs*. Translated by Marina Ramsay. Edited by Richard Pierce. Kingston, Ontario, Canada: The Limestone Press.

Kutchin, Howard M. 1898. *Report on the Salmon Fisheries in Alaska, 1898*. Washington DC: GPO.

Kutchin, M. "Fox Breeding on the Alaska Islands." *Scientific American* 82 (April 1900): 242.

LaGuardia, Fiorello, "Minority Report, Leasing Public Lands in Alaska for Fur Farming," Committee on the Public Lands, H.R. 678, 69th Congress, Session I, March 27, 1926.

Leekley, James R. "Petersburg Station Seeks to Establish Standards For Fur Farming Industry," *Fairbanks Daily News-Miner*, September 9, 1943.

Lloyd-Jones, Alice. 1918. "The Fur Industry." Master's thesis, University of Wisconsin.

Loftus, Jule B. ca. 1970. Untitled memoir. Unpublished manuscript, n.p. In author's possession.

Mangusso, Mary Childers, and Stephen W. Haycox. 1989. *Interpreting Alaska's History: An Anthology*. Anchorage: Alaska Pacific University Press.

Masterson, John, and Helen Brower. 1948. *Bering's Successors: 1745–1780*. Seattle: University of Washington Press.

McAllister, Helen Judge. 1943. "Memoirs of Pribilof Islands: Nineteen Years on St. George and St. Paul Islands, Alaska." Manuscript, n.d. Helen Judge McAllister Papers. University of Washington Libraries, Special Collections OCLC# 123949607.

McGowan, Sarah. "Fox Farming: A History of the Industry in the American Period." In *The History and Ethnohistory of the Aleutians East Borough* by Lydia Black, 244–251. Fairbanks, AK: The Limestone Press, 1999.

Miller, Char. 2001. *Gifford Pinchot and the Making of Modern Environmentalism.* Washington, DC: Island Press.

Miller, S.V.B. "Why Cats Are Wild." *The Fur Farmer Magazine* 1, no. 9 (December 1923): 10.

Morgan, Lael, ed. 1992. *The Aleutians. Alaska Geographic* 7, no. 3. Anchorage: Alaska Geographic Society.

Murie, Olaus J. 1959. *Fauna of the Aleutian Islands and Alaska Peninsula*, North American Fauna Number 61. Washington DC: U.S. Fish and Wildlife Service.

Niedieck, Paul. 1909. *Cruises in the Bering Sea*, trans. R.A. Ploetz. New York: Charles Scribner's Sons.

Novak, Milan et al. 1987. *Wild Furbearer Management and Conservation in North America.* Ontario, Canada: Canadian Ministry of Natural Resources.

Oleksa, Michael. 1992. *Orthodox Alaska: A Theology of Mission.* Crestwood, NY: Saint Vladimir's Seminary Press.

Osgood, Wilfred H. 1908. *Silver Fox Farming.* USDA Farmers' Bulletin 328. Washington, DC: GPO.

Osgood, Wilfred H., Edward A. Preble, and George H. Parker. 1915. *The Fur Seals and Other Life of the Pribilof Islands, Alaska, in 1914.* Washington, DC: GPO.

Paul, Fred. 2003. *Then Fight For It!* Victoria, B.C., Canada: Trafford Publishing.

Paul, Thomas W. 2009. *Game Transplants in Alaska.* Technical Bulletin #4, second edition, Juneau, AK: Alaska Department of Fish and Game.

Parker, Margery Pritchard. "A Northern Crusoe's Island," *National Geographic* 44, no. 3 (September 1923): 313–326.

Pedersen, Christian T. Christian T. Pedersen Papers 1920–1966, Consortium Library Archives, Anchorage.

Pederson, Susie. "Ed Ophiem, Sr. 'Fox Farming,'" *Elwani* 6 (1979): 5–9.

Pierce, Richard A. 1990. *Russian America: A Biographical Dictionary.* Kingston, Ontario, Canada: The Limestone Press.

Pinchot, Gifford. 1910. *The Fight for Conservation.* New York: Doubleday, Page, & Company.

Preble, Edward A., and W. L. McAtee. 1923. *A Biological Survey of the Pribilof Islands, Alaska.* North American Fauna, No. 46, Part 1. Washington DC: GPO.

Rakestraw, Lawrence. 1994. *A History of the United States Forest Service in Alaska.* 1981. Reprint, Anchorage, AK: USDA Forest Service.

Redington, Paul. "Fur Farming." *The Fur Farmer Magazine* 7, no. 7 (January 1932): 4–5, 8.

Roberts, Larry D. "A Preliminary Survey of Historic Southeastern Alaskan Fur Farming." (Working paper supplied by author, February 2010).

Robey, Ralph. "Fur Coats and Winning the War," *The Fur Journal* 9, no. 1 (November 1942): 15.

Rocking Moon. Directed by George Melford. Metropolitan Picture Corporation of California, 1926.

Romance of Furs, The. Washington, DC: Woodward and Lothrop, 1936.

Rossiter, Harriet. "Alaska's Fur Industry, Its Decline and Rise." *Fur Trade Review*, (January 1923), 173–176.

Samet, Arthur. 1950. *Pictorial Encyclopedia of Furs*. New York: Arthur Samet Book Division.

Sather, Josephine. "Fox Farm at Nuka Bay," four part series. *The Alaska Sportsman*, "The Island" 12, no. 7 (July 1946); "The Foxes" 12, no. 8 (August 1946); "The Birds and the Bears" 12, no 9. (September 1946); "Our Glorious World" 12, no. 10 (October 1946).

Scharfenberg, Joni. "Fur Farming in Alaska," *Alaska Farm Magazine* 2, no. 1 (January 1982): 36–37.

Schlung, Tyler M., and Nikolski students 2003. *Umnak: The People Remember*. Walnut Creek, CA: Hardscratch Press.

Schrader, Barbara, and Paul Hennon. *Assessment of Invasive Species in Alaska and Its National Forests*. www.akweeds.uaa.alaska.edu/pdfs/literature/Schrader_R10_Inv_Spp_Assessment.pdf

Seton, Ernest Thompson. 1909. *Biography of a Silver Fox*. New York: The Century Company.

———. "Raising Fur-Bearing Animals for Profit," *Country Life in America* 9 (January 1906): 294.

"Seven Ways to Compute the Relative Value of a U.S. Dollar Amount, 1774 to Present." www.measuringworth.com.

Shade, Charles I. 1949. "Ethnological Notes on the Aleuts." Honors thesis, Harvard University.

Shillinger, J. E. "Humane Aspects in Fox Farming," speech, American Society for Prevention of Cruelty to Animals, St. Louis, MO, September 30, 1929.

Smith, J. L. 2001. *Russians in the Pribilof Islands 1786–1867*. Anchorage, AK: White Stone Press.

Snider, Gerrit. 1929. *Mink Raising in Alaska*. Anchorage, AK: The Times Publishing Company.

Sowls, Arthur, Scott A. Hatch, and Calvin J. Lensink. 1978. *Catalog of Alaskan Seabird Colonies*. Washington, DC: Biological Services Program, U.S. Fish and Wildlife Service.

Spector, Robert. 1998. *Seattle Fur Exchange: 100 Years*. Seattle, WA: Documentary Book Publishers.

Stejneger, Leonhard. 1936. *Georg Wilhelm Steller: The Pioneer of Alaskan Natural History*. Cambridge, MA: Harvard University Press.

Sterling, Keir B. "Builders of the U.S. Biological Survey, 1885–1930." *Journal of Forest History* (October 1989): 180–187.

Strohmeyer, John. 1993. *Extreme Conditions: Big Oil and the Transformation of Alaska*. New York: Simon & Schuster.

Stubbs, D. E. "About Worms That Infest Fur Bearers." *Seward Daily Gateway*, October 23, 1931.

Stuck, Hudson. 1917. *Voyages on the Yukon and Its Tributaries*. New York: Charles Scribner's Sons.

Swanson, Henry, and high school students at Unalaska. 1982. *The Unknown Islands: Life and Tales of Henry Swanson*, Cuttlefish IV. Unalaska: Unalaska City School District.

Taylor, Hilda Harbin. 1994. "Say That Reminds Me!" (manuscript). MS 168, Alaska State Historical Library, Juneau.

Tewkesbury, David, and William Tewkesbury. 1947. *Tewkesbury's Who's Who in Alaska and Alaska Business Index*. Juneau, AK: Tewkesbury Publishers.

Thompson, Seton H. *Alaska Fishery and Fur-Seal Industries, . . .* [titled by year 1947–1957]. Washington, DC: GPO, years 1948–1958.

Tikhmenev, P. A. 1978. *A History of the Russian-American Company*, trans. and ed. Richard A. Pierce and Alton S. Donnelly. Seattle: University of Washington Press.

Tingle, George R. 1897. *Report on the Salmon Fisheries in Alaska 1896*. Washington, DC: GPO.

Torrey, Barbara Boyle. 1983. *Slaves of the Harvest: The Story of the Pribilof Aleuts*. St. Paul Island, AK: Tanadgusix Corporation.

"Transcript of Notes from St. Paul Island, 1872–1909." Compiled by Walter L. Hahn 1910–1911. U.S. Fish and Wildlife Collection, MS 51, Box 2, Folder 7. Alaska Historical Library, Juneau.

Travis, Hugh, and J.R. Leekley, et. al. 1972. *An Interdisciplinary Study of the Effects of Real and Simulated Sonic Booms on Farm-Raised Mink*. Beltsville, MD: U.S. Agricultural Research Service.

Trut, Lyudmila N. "Early Canid Domestication: The Farm-Fox Experiment." *American Scientist*, 87 (March–April, 1999): 160–169.

Turner, L. M. 1886. *Contributions to the Natural History of Alaska*. Washington, DC: GPO.

Unalaska City School District. 1986. *People of the Aleutian Islands*. Unalaska, AK: Unalaska City School District.

Underwood, John J. 1913. *Alaska: An Empire in the Making*. New York: Dodd, Mead and Company.

Underwood, Larry S., and Jean Baldrige. 1981. *Feasibility of Fox Farming in the NANA Region*. Fairbanks: Arctic Environmental Information and Data Center.

U.S. Bureau of Biological Survey, Department of Agriculture. "Island Blue Fox Ranchers of Alaska: Stocked January 1, 1923." Press release, June 1923.

———. "Shipments of Furs from Alaska from . . . to . . . ." Annual press releases 1919–1923.

U.S. Bureau of the Census, Department of Commerce. 1932. *Fifteenth Census of the United States, 1930, Outlying Territories and Possessions.* Washington, DC: GPO.

U.S. Bureau of the Census, Department of Commerce. *Nonpopulation Census Schedules for Alaska, 1929: Agriculture.* Record Group 29, National Archives and Record Administration, Pacific Alaska Region, Anchorage.

U.S. Bureau of Fisheries, Department of Commerce. *Fox Farming in Alaska*, Bureau of Fisheries Document 834. Reprinted in *Fur News* (June 1917): 16.

U.S. Forest Service, Department of Agriculture. Land Case Files 1920–1974 and 1907–1989. Record Group 95, National Archives and Records Administration, Pacific Alaska Region, Anchorage.

———. Records of the Forest Service, 1870–2000. Record Group 95, National Archives and Records Administration, Pacific Alaska Region, Anchorage.

U.S. General Land Office, Department of the Interior. "Register of Final Homestead Receipts and Fur Farm Rentals, July 1, 1899 to March 2, 1931." National Archives and Records Administration, Pacific Alaska Region, Anchorage.

Veniaminov, Ivan. 1984. *Notes on the Islands of the Unalashka District.* Translated by Lydia T. Black and R. H. Geoghegan. Kingston, Ontario: Limestone Press.

A Village Fur Farm. Fairbanks: Interior Village Association Alternative Development Program, 1982.

Walker, Lillian. Interview by Debbie Turner and Karen Brewster September 2, 2003. University of Alaska Oral History Project, H2002-10-09, Part I.

Washburn, M. L. 1901. "Fox Farming in Alaska." In *Harriman Alaska Expedition*, Vol. 2, by William H. Dall et al., 357–365. New York: Doubleday, Page & Company.

Willoughby, Barrett. 1925. *Rocking Moon.* New York: G.P. Putnam's Sons.

Yearbook of Agriculture, 1937. Washington, DC: GPO, 1937.

Zimmerman, C. E. "Domesticating Alaska Blue Foxes." *The Pathfinder of Alaska* 4, no. 11 (September 1923): 1–3.

Periodicals Cited

Alaska Farm and Garden
The Alaska Sportsman
Alaska-Yukon Magazine
American Fox and Fur Farmer
American Fur Breeder
American National Fur & Market Journal

The Fox Breeders Gazette
The Fur Farmer Magazine
Fur-Fish-Game
Fur Farmers' Bulletin
The Fur Journal
Fur News
Fur Trade Review
Journal of the American Veterinary Medical Association
Journal of Forest History
Journal of Mammalogy
National Geographic
The Pathfinder of Alaska
Popular Science
Science
Scientific American

Newspapers Cited

The Alaska Appeal
The Alaska Citizen
Alaska Daily Dispatch
The Alaska Weekly
Anchorage Daily Times
The Auburn Citizen (New York)
Cordova Daily Herald
Fairbanks Daily News-Miner
Juneau Empire
The Mining Journal (Ketchikan)
The Nenana Daily News
Nome Nugget
Oswego Daily Palladium (New York)
Petersburg Herald
Petersburg Press
Seattle Times
Seward Weekly Gateway
Skagway Daily Alaskan
The Tacoma Times
The Valdez Miner
The Wrangell Sentinel

INDEX

A

Abbes, Bill, 124, 134–135
Adak Island, 158
Adak Ranching Company, 145–146
Adler, Don, 111, 135
Admiralty Island, 147
Afognak Island, 97
Agattu Island, 52, 159
Agattu Natives, 12
Agricultural Experiment Station, 131
Air Force, 173
Alaska and Its Resources (Dall), 21
Alaska Blue Fox Farmers Association, 81, 83
Alaska Commercial Company, 23, 25, 38, 58
Alaska Development Board, 157–158, 164
Alaska District Court, 170
Alaska Experimental Fur Farm, 115–118, 128–132, *131, 152*–154, *153,* 169–173, *172*
Alaska Farm Magazine, 176
Alaska Fur and Silver Fox Company, 41–42
Alaska Fur Farm Leasing Act, 96–97
Alaska Game Commission, 94, 110–114, 118, 129, 133, 141, 146, 152, 154, 168
Alaska Glacier Seafood Company, 129
Alaska Native Brotherhood, 96–97
Alaska Natives, 95–97, 161–162, 164, 177. *See also* Aleuts
Alaska Railroad, 69, 105, 121, 125, 135
Alaska Renewable Resources Corporation (ARRC), 175
"Alaska sable." *See* skunk farming
The Alaska Weekly, 68
Alaska Western Fur Corporation, 107
Alaskan Fur Corporation, 135
Albery, O. D., 113
Alcan Highway, 141
alcohol, 109, 126, 177
Aleutian Fur Company, 159
Aleutian Island Wildlife Refuge, 49, 54, 95, 145, 158–160
Aleutian Islands, 7–21, 19, 42, 49, 51–54, 70, 141, 144–147, 158–160
Aleutian Islands Fur Farmers Association, 135, 145–146
Aleuts, 12–13, 18, 20–21, 25–26, 50–54, 62, 146–148, 161
Alexander Archipelago Forest Reserve, 43
All-Alaska Sweepstakes, 70
Amchitka Island, 159
American National Fox Breeders Association, 84
American Society for the Prevention of Cruelty to Animals, 174
Ames, Jess, 169

Index

Amiq Island, 17
Amundsen, Roald, 69
Anchorage Agricultural Fair, 116, *117*
Anderes, Ernest, 168–169
Andreanof Islands, 8, 127
animal rights activists, 173–176
Applegate, Samuel, 33, 50–54, 61–62
Ashbrook, F. G., *82,* 83, 100, 106, 148, 160, 174
Attu Island, 12–14, 16, 52, 54, 147
Attu Native Community, 145
auctions, 28, 66, 81, 132, 149, 167, 175
automobiles, *67*

B

backyard farms, 111–112
Barker, Kay, 84, 88
bears, 48–49
beaver farming, 72
Bellanca plane, 121–*122*
Belyaev, Dimitry, 91
Bennett, A. W., 133, 135–136
Bennett, Amanda, 84
Bergmann, Harold and Ethel, 169
Bering, Vitus, 8, 11
Bering Sea, 70
Berkh, Vasilii, 20
Berry, Forest, 90–91
bird colonies, 17, 20–21, 49–50, 55, 144–146, 159, 177. *See also* eagles
The Black Fox Magazine, 119
"black marten." *See* skunk farming
Blakely, Rufus, 47
Blatchford, Percy, 138
Blue Fox Farming in Alaska (Walker), 100–101
"Blue-Fox Farming in Alaska," 83
Bolshanin, Nicholas, 52–54, *53,* 64, 71–72, 74, 135, 159, 176
"Bone Dry Law," 109
bounty, 1, 20, 54–57, 78, 126, 145

Bowman, Harold, 127–128, 146, 159
branding. *See* Fox Branding Law
Brothers Island, 79, 89, *90,* 133, 135
Buldir Island, 145
Bunnell, Charles, 121, 128, 131
Bureau of Biological Survey, 42, 45–46, 52, 54, 57, 62, 68, 70, 115, 128–129, 133
Bureau of Customs, 107
Bureau of Fisheries, 62–63, 145
Bureau of Indian Affairs, 169

C

C. M. Lampson & Company, 28, 62
Cadzow, Daniel and Rachel, 69
Canadian farmers, 119
Canadian National Express, 108
Carlson, Otto, 38
Catherine the Great, 20
census, 84, 111–113, 157
A Century of Servitude (Jones), 25
Chevigny, Hector, 20
Chief Titus, 39
chinchilla farming, 165–166
chinchilla rabbits, 111, 116
Chinese merchants, 7–8
Christensen, Oscar, 125
Christie, Irene and "Rusty," 175–176
Chugach National Forest, 44–*45,* 49, 51, 68, 120, 124, 129, 136, 168
Civilian Conservation Corps, 128–131
Cleary, Grover, 107
closing farms, 118–119, 161–162, 164
clothing, 63, 119
  fox stole, *29,* 30
  luxury, 177
  mink coats, 105
  movie stars and, 166–167, 174
  muskrat fur used for, 148
  ostrich-feather hat, *29*
  postwar styles of, 160, 164–168

short-haired pelts, 174
silent films and, 77
silver fox scarves, 106
synthetic materials used for, 141, 148–149, 164–165, 175
utilitarian, 67, 149, 164–165, 174, 177
C&M Fur Farm, 176
Coast Guard, 79
coastal farms, 40, 42, 45, 94, 125, 126, 154
Cold War, 167
Commander Islands, 7, 16
commercial loans, 119
Conclusion Island, 107
Cook Inlet and the Susitna Mink Breeders Association, 116
Cooks Inlet Blue and Black Fox Farmers Association, 83
Coombs, Clyde, *82*
Cooperative Extension Service, 128
Copper Center Agricultural Experiment Station, 40–42, 64

D
Dall, William, 21
Dalton, Charles, 38
Dean, Ballard, 128
DeGama Land, 8, 20
Delarof, 147
Department of Agriculture, 41, 50, 51, 52
Department of Animal Husbandry, 114
Department of Commerce, 50–51, 60–62. *See also* Bureau of Fisheries
Department of the Interior, 95, 145, 146
deportation, 144
DeVenne, George, 158–159

Dimond, Tony, 146
Division of Economic Ornithology and Mammalogy. *See* Bureau of Biological Survey
Dufresne, Frank, 70, 129
Dugdell, W. H., *44*

E
eagles, 54–58, *56*
"Easy to Raise, Hard to Catch," 57–58
Eckardt, Hugo, 121–122
ecological zones, 46
The Economics of the Fur Industry, 167
El Capitan Island, 133
Eleanor Island, 102
EMBA (Mutation Mink Breeders Association), 167, 175
endangered animals, 55, 147, 159
Erskine, W. J., 58–61, *59*, 64, 89, 92, 135
Eskimo (film), 134
Etolin Island, 47
experimental farms. *See* specific farms
extinction, 46, 174

F
Fairbanks Daily News-Miner, 96, 113, 128
Fanning, John, 47
Farm Credit Administration, 119
farmers. *See* specific farmers
fashion. *See* clothing
Fassett, Harry C., 62
Faulconer, Katherine "Kitty," 109
feed house/traps, *31*
fences, 19
Fish and Wildlife Service, 133, 145–148, 159–160
fishermen, 35, 79–80, 142
fitch, 132–133

Index

Forbes, William, 48, 176
Foreign Claims Settlement Commission, 159
Forest Service, 43–45, 51, 68, 96, 120, 129, 134–135, 136, 161–162, 164
Fouke Fur Company, 148
Fox Branding Law, 79–81, 111, 113
Fox Breeders Protective Association of Alaska, 27
fox circuses, 90–91
"Fox Farming in Alaska" (Washburn), 24, 27, 32
Fox Islands, 14, 15, 126
"Fox Ranchers Getting Rich," 78
foxes. *See also* pelts
   arctic fox, 7, 9, *10–11*
   blue fox, 8, 11–12, *15*–18, 20, *33*, 70, 88, 124, *153*
   branding, 79–81, 111, 113
   breeders, 32–33, 36–38, 61, 71, 78, 88, 91, 105–108, 116, 131–132
   captured, 27
   diseases of, 60, 92–94, 97, 102, 108–109, 113, 124, 132
   domestication of, 2, 36, 89–92
   eradication of, 158–160, 164
   farming of, 23, 24, 27–28, 41–42, 79–81, 105–107
   food of, 30–33, 40, 57, 60, 65, 70–71, *86*, 92–94, *103*, 105, 119, *125–126*, 132–135, 144, 152, 154, 158, 161
   free-range, 38–39
   Greenlandic, 34–36
   inbreeding of, 18, 64, 164, 177
   negligence of, 108–109, 147
   population of, 25, 30–33, 112
   prey of, 12
   red fox, 14–15, 38
   Russian fox, 92
   Siberian fox, 12–14
   silver fox, 14–16, *39*, 69, 100–101, *112*, 173–176
   silver-black fox, *14*
   transplant of, 18–19, 20–21
   value of, 30, 41
   white fox, 32, 33, 70, 116
Freuchen, Peter, 134
Fromm Brothers, 107, 121
Funsten & Bros., 62
Funter Bay, 147
Fur Breeders Agricultural Cooperative, 169–170
fur breeders fair, 116
fur farm associations. *See* specific associations
The Fur Farmer Magazine, 74–75, 78, 83–84, 113, 116–118, 127
Fur Farming (Harding), 46
Fur Farming Opportunities in Alaska, 159, 164, 168
Fur Products Labeling Act, 165
fur seals, 16–17, 24–26, 31, 40–41, 62, 115–116, 148
The Fur Trade Journal, 128
furs. *See* pelts

G

game wardens, 81, 111
Garfield, Charles, 76
General Land Office, 95, 111, 113, 120
George W. Elder, 23–24
Glotov, Stepan, 15, 20
gold, 35, 67, 128, 178
Goshaw, George, 70, 121, *122, 124*, 129, 138, 155, 160
government. *See also specific departments*
   aid from, 119–120, 136
   control from, 149–150
   farms of, 92, 127, 138, 177

leases from, 26–27, 50–51, 61, 70–71, 80, 94–95, 146
licenses from, 60, 73, 111, 133–135, 141
permits from, 44–45, 49, 52, 54, 68–69, 94, 127, 134, 146, 162, 168
propagation permits from, 36–37, 72, 111–113
sales from, 24–27
threats from, 144
Graves, Earl, 101–*103*, 105–110, 113–114, 139–140
Graves, Kitty, *110*, 114, *171*
Gray, Billy, 81
Great Depression, 115–140
Grier, Ben, 101–102
Gross, A. C., 111
Grosvold, Andrew, *99*, 111
Gruening, Ernest, 146
Gunner, 126

H
habitat loss, 30, 46
Hadlund Farm, *86*
Haida Indians, 96
Harbor Island, 162
Harding, A. R., 46
Harriman, Edward, 23–24
Harriman Alaska Expedition, 23–24, 27, 43, 46
Haynes, Ole, 81, 83
Hegness, John, 70, 134
Heideman, Charles W. H., 40–42
Heintzelman, B. Frank, 129, 162, 164
Hesketh Island, 34
Holbrook, Wellman, 129
Homestead Act, 26, 95
hostages of Russian explorers, 12
House Committee on Public Lands, 95

Hudson's Bay Company, 28, 175
"Huge Profits in Fox Farms," 78
hunting bans, 50
Hyperborean Ranch, 70

I
Ibach, Joe and Caroline ("Muz"), 65, 68–69, 136, *137*
Ickes, Harold, 145–146
Igadik, 17
Illustrated Catalogue of Fur Bearing Animals, 165
immigrants, 51, 112, 142
imports, 139, 164
Indian Claims Commission Act, 162
influenza, 66
interior farms, 40, 69, 94, 107, 125, 154, 170
internment, 142, 144
island farms. *See* specific islands

J
Japanese, 79, 141–148
Judge, James, 30–33, 92, 176

K
Kachemak Bay, 34
Kalvaga Island, 54
Kanaga Ranching Company, 127, 146
Kanak Island, 162
Kenai Peninsula, *85,* 135
Ketchikan, 162, *168*
Khlebnikov, Kiril, 19
Kiska Island, 147
Knight Island, *110*
Kodiak Fur Farm, 58, 60–61, 88–89, 92
Kodiak Island, 19
Korean Peninsula, 142

L

labor shortages, 139
LaGuardia, Fiorello, 96
land conservation, 19, 43–44. *See also* wildlife refuges
Land Office, 69, 101
land rights, 94
Lane, Louis, 126
Larianoff, Sasha, 27, 88–89
Lebedev-Lastochkin Company, 17
Leekley, James, 140, 150–152, *151*, 154, 170, *172*, 173
Lemesurier Island, 69, 136
Leviathan, 106
Liljegren, Fred, 34–35, 64, 84
Lind, Knute, 111, 135
Lloyd, T. W., 74
Loftus, Jule, 120–125, *122, 123*, 128–129, 131, 139–140
Long Island, 133. *See also* Semidi Fox Propagation Company
Long Island Farm, 27–28, 32, 58, 60–61, 109, 135–136
losses and profits, 63–64
Louisiana Purchase, 76

M

magazines, 166–167
malnutrition, 147
markets, 92–94, 139–140, 178
marmots, 113
marten farming, 47, 72
Matanuska Colony Project, 128
Matanuska Minkery, 74
Maycock, Bert, 47
McCarthy, Joseph, 167
McCrary, Nelson, 133
McCullough, Kenneth, 107
measles, 147
Mendenhall Valley, *163*
merchants, 73–79
Merriam, C. Hart, 46
MGM Films, 134
Middleton Island, 33–35, 65, 68
Midway Island, 134
Migratory Bird Act, 55
military draft, 141–142, 150
Milligan, J. Edward, 64, 107–108, 134
miners, 67, 95, 109, 112, 142
mink
  farming of, 47, 72, *104*–106, 115, 125, 128, 136, 138, 142–*143*, 154, 164, 166–170, 176
  mink depression, 173–174
  mutations, 166, 173–176
  noise damages and, 170–173
Mink Raising (Ashbrook), 106
Mink Raising in Alaska (Snider), 105–106
M&M Farms, 107–108, 121, 134
monetary systems, furs used for, 8
A Monograph of the Silver Fox (Heideman), 41
monopolies, 18
Monte Carlo Island, *86*
moonshine, 109, 126
Morgan, Thomas F., 24
Morrison, George L., 39–40, 64, 107–108, 134
movie stars, 166–167, 174
Mt. McKinley National Park, 170, *171*
Muir, John, 43, 44
Murie, Olaus, 121, 133, 144–146
muskrat farming, 47–48, 72, *99*, 133
MV *Sea Otter*, 83

N

National Association of the Fur Industry, 118
National Experimental Fur Farm, 100–102, 121, 140
National Geographic, 68

national land, 43–45, 94–95
National Reconstruction Administration, 119
National Zoo, 100
Native Rights, 161–162, 164
natives. *See* Alaska Natives
naturalization, 142, 144
Nenana News, 73
Noatak River, 121–122
noise damages, 170–173
non-native civilians, 71, 145, 147, 157
non-native species, 42, 49–50, 133
North Semidi Island, 24, 57
Northern Commercial Company, 58, 60, 76
Nuka Island, *87,* 88, 155
Nunivak Island, 169
nutria, 133

O

Office of Indian Affairs, 146, 161
Office of Price Administration, 150
Ohmer, Earl, 101, 129, *130,* 138, 155, 168–169, 176
Ollestad, Tallak, 126
Opheim, Ed, 79, 109, 170
Ornithologists Union, 55
Osgood, Wilfred, 46
ostrich feathers, *29,* 63
Oulton, Robert, 38

P

Parker, Basal, 97–101, *98,* 134
Parks, George A., 101–102, 109, 113–114, 120
The Pathfinder of Alaska, 89–90, 135
Patterson Island, 34
Paul, William, 96
Pearl Harbor, 140–144
Pedersen, C. T., 159
pelts, 9–10, *93. See also* clothing; foxes; mink
  blue fox, *45,* 65–66, 72, *93,* 115, 127, 132, 174
  cost of, 28, 61–66, 70–71, 74, 92–94, 105–108, 115, 117, 125, 127, 132, 149–150, 160, 164–165, 169, 175
  demand for, 8, 117
  importation of, 139, 164
  improvement of, 160–161
  mink, 132
  monetary systems and, 8
  platinum blue, 154
  regulations on, 25
  silver fox, 66, 93, 115, 132, 149, 174
  using dye or bleach on, 16, 36, 48, 70–71, 78, 165
  utilitarian, 67, 149, 164–165, 174
pens, 36–40, *39,* 60, 69–70, 107, 113, 129, *131,* 132, *168,* 169
Perl Island, *82*
PETA, 173–176
Petersburg Farm. *See* Alaska Experimental Fur Farm
Philadelphia Ledger, 35
Pierce, Richard, 20
Pinchot, Gifford, 43–44
Piper, A. F., 72
poachers, 19, 26, 35, 47, 57, 79–81, 92, 94, 134
Polet, Alvin, *122*
porcupines, 125, 135
porpoises, *126*
predatory animals, 47, 55
Pribilof Fox Farm, 30–33
Pribilof Islands, *10–11,* 16–19, 25, 26, 40–41, 50, 62–63, 71, 92–*93,* 115–116, 138, 147–148, 160–161
Pribylov, Gavriil, 17–18, 20

Prickly Heat, 99
Prince Edward Island, 38, 64, 107–108, 121
Prince William Sound, 65–66, 84
Prince William Sound Fox Farmers Association, 83
profits and losses, 63–64
Prohibition, 109, 126
property rights, 146
protected land. *See* wildlife refuges
Public Works Administration, 128

Q

Queen Mary, 119

R

raccoon farming, 72, 133
Race, Harry, 108–109
Rat Islands, 19
Reagan, Nancy, 176
Redington, Paul, 118
Redpath, James, 24
reindeer herds, 100
reports, 102, 105, 107, 109, 113–114, 124, 133
reservation land, 25–26
Rice Fur Farm, 175–176
Richardson Road, 105, 107
Riggs, Thomas, Jr., 66
Rocking Moon, 27, 88–89
Rocking Moon Island, 88–89
Roosevelt, Eleanor, 119
Roosevelt, Franklin, 128, 136
Roosevelt, Theodore, 42–44, 49
Ross, Victor, 121, *122*
Round Island, 162
Rudy, Charles, *163*
Rudy's Ranch, *163*
Russian explorers, hostages of, 12
Russian-American Company, 18–20, 21, 25, 58

S

salmon, 178
salted meat, 31–32, 125
Samalga Island, 33
San Juan Fox Farm, 81
Sand Point, *99*
Saratoga Springs Farm, 100–102, 121, 140
Sather, Josephine, 65, 84, *87,* 88, 155
scams, 36
Schove, George, 79
Scientific American, 48
Scott, Louis, 133
Scove, Mrs. George, 34
sea otters, 8–9, 50
Seattle Box Company, 107
Seattle Fur Exchange, 73–79, *77, 78,* 116, 118, *138,* 167, 175–176
Semidi Fox Propagation Company, 26–33, 38, 42, 57–58, 99, 109
Seton, Ernest Thompson, 67–68, 94
Seward Daily Gateway, 170
Seward Peninsula, 138
Sheely, Ross, 128
sheep, 128
Sheldon, Bobby, *67*
shipping rates, 167
Shishmaref, 121–122
Shitiki, 9
Sholtz, Rupert, 135
Signal Corps, 50
silent films, 77
Silver Fox Farming (Ashbrook), 100
Silver Fox Farming (Osgood), 46
Singa Island, 133
Sinnott, Nicholas, 95
Sister's Island, 81
Skagway Daily Alaskan, 48
skunk farming, 48, 72, 176
Slaves of the Harvest (Torrey), 25
Smith, Sumner, 101–102

Smith, Thomas Vesey, 33–35, *37,* 65, 79
Smithsonian Institution, 40
Snider, Gerrit, 105–*106,* 129, 138, 155, 176
snowshoe hares, 125
sonic booms, 170–173, *172*
Southeastern Alaska Fox and Fur Farmers Association, 83, 92
Southwestern Alaska Blue Fox Farmers' Association, 100
Spanish Flu Pandemic, 66
squatters, 69, 96
squirrels, 19
SS *Admiral Farragut,* 66
SS *Alaska,* 107–108
SS *Leah,* 41
SS *Victoria,* 70
St. George Island, 17–20, 30–33, 92, 161
St. Paul Island, 18–20
statistics, 110–114
Stefansson, Vilhjalmur, 69
Steller, Georg, 8
Storey Island, 34, 35
Storm Island, 134–135
The Story of the Silver Fox (Heideman), 41–42
Stubbs, D. E. "Duke," 170–*171*
Stuck, Hudson, 39
Sukoi Island, *39,* 79
Sullivan Island, *126*
Sumdum Island, 34
Sutherland, Dan, 96
Sv. Georgii Pobedonoset (St. George the Victorious), 17
Sv. Ioann (St. John), 7, 9

T
Taft, William, 49
Taigud Island, 89
Tanaga Island, 127
Tanana Valley Fair, 116
tariffs, 119, 132
Tashman, Lilyan, 88
taxes, 73–79, 83, 119–120, 128, 149–150, 167
Taylor, W. B. "Preach," 24
Tazlina River Farm, 40
Temple, P. R., 35
Territorial Department of Audit, 80
Teton Fox Farm, 74
Tolovana Hot Springs, 39, 107–108
Tolstykh, Andrean, 7–13, 20, 176
Tongass National Forest, 44, 49, 51, 68, 95–96, 120, 124, 136, 168
transportation routes, 69
traps and trapping, 13, 19, 30–33, *31, 37,* 161
Treasury Department, 25–26, 30, 44, 50
Trefzgar, Hardy, *44*
trespassing signs, 80, 161–162
Troy, John, 129
tuberculosis, 97, 102, 121, 129, 147
Tunulgasen (chief of Attu), 13
Turnbow, Guy, *103,* 107, 109, 135
Turner, Lucien, 144

U
Umnak Island, 71, 126–127
Umnak Native, 71, 126
Unalaska, 51–52, 147
Unalga Island, 52
Unimak Island, 17, 20
"uninhabited" islands, 26–27
United States government. *See* government
University of Alaska, 152, 154
Urata, Ryotaro and Chiyo, 142–144, *143*
Ushagat Island, 88

V

Valdez Glacier, 35
Valdez-Fairbanks Road, 69
Van Scheel, Bob, 134
Vasil'ev, Ivan, 21
veterinarians, 94, 97–*103,* 105, 109, 113–114, 120–125, *123,* 128, 133, 139, 164
Virginia IV, 102
voting rights, 84

W

Walker, Ernest P., 83, 100–101, 110
Wallen, Rika, 84
War Manpower Commission (WMC), 150
Warren Island, 133
Warsaw Zoo (Germany), 148
Washburn, M. L., 21, 23–33, 38, 58, 176
"Wealth in Alaska Foxes," 78
weather stations, 50
Webster, F. C., 72
Wei, Peter, 109
Weschenfelder, Eugene and Marie, *153*
Whale Island, 57, 97–99, *98,* 134
white entitlement, 161–162
Whitestone Farms, 176
wildlife refuges, 49, 177
Williamson, Billy and Harriet "Mickey," 84, *85,* 108, 135
Williamson Silver Black Fox Ranch, *85, 91, 117*
Willoughby, Barrett, 88–89
Windfall Island, 47
women, 84–89, 142
World War I, 57, 62, 63–64, 66, 76, 79–80
World War II, 135, 141–155
The Wrangell Sentinel, 142

Y

York, James, 34, 64
Yukon Fur Farm, 129, 138, 168–169
Yukon Island, 34

Z

Zimmerman, C. E., 89–91, *90,* 133, 135